The Cattleyas and Their Relatives
The Debatable Epidendrums

The Cattleyas and Their Relatives
The Debatable Epidendrums

CARL L. WITHNER
AND
PATRICIA A. HARDING

TIMBER PRESS
Portland • Cambridge

For all those orchid hobbyists and researchers
who have puzzled over the correct names
for these pseudobulbous species
originally called *Epidendrum*

Copyright © 2004 by Carl L. Withner and Patricia A. Harding
All rights reserved.

Published in 2004 by

Timber Press, Inc.
The Haseltine Building
133 S.W. Second Avenue, Suite 450
Portland, Oregon 97204, U.S.A.

Timber Press
2 Station Road
Swavesey
Cambridge CB4 5QJ, U.K.

www.timberpress.com

Printed through Colorcraft Ltd., Hong Kong

Library of Congress Cataloging-in-Publication Data

Withner, Carl L. (Carl Leslie)
  The Cattleyas and their relatives : the debatable Epidendrums / Carl L. Withner and
    Patricia A. Harding.
      p. cm.
    Includes bibliographical references and index.
    ISBN 0-88192-621-3 (Hardcover)
    1. Orchids. 2. Cattleyas. 3. Epidendrum—Identification. I. Harding, Patricia A. II. Title.
  SB409 .W7934 2004
  635.9'344–dc21
                                                                                    2003013816

A catalog record for this book is also available from the British Library.

Table of Contents

Preface   7

Introduction   13
    Lindley's Key (1853) to the Twelve Subgenera of
        Epidendrum   20
    Key to the Genera in This Book   23

CHAPTER 1   The Genus *Anacheilium*   27

CHAPTER 2   The Genus *Coilostylis*   137

CHAPTER 3   The Genus *Encyclia*   148

CHAPTER 4   The Genus *Hormidium*   152

Gallery of Debatable Epidendrums   161

CHAPTER 5   The Genus *Oestlundia*   201

CHAPTER 6   The Genus *Panarica*   207

CHAPTER 7   The Genus *Pollardia*   216

CHAPTER 8   The Genus *Prosthechea*   251

CHAPTER 9   Miscellaneous Small Genera and Aberrant Species   278

Selected References for Additional Reading   285

Index of Persons   289

Index of Plant Names   290

Preface

Six volumes of *The Cattleyas and Their Relatives* have apparently not been enough for some people. We have repeatedly been asked why certain *Epidendrum* species were not included in those texts. The answer is, of course, that they were not cattleyas, laelias, or encyclias. In this book we will deal with more of the *Epidendrum* species, especially those that have fallen into *Encyclia* in the past. Some of these original epidendrums have been assigned to *Encyclia* more or less only because they had pseudobulbs, and because other salient characteristics of the plants and flowers were ignored. Perhaps such qualities also did not rank highly enough in various estimations to be considered as significant in producing a "robust" classification system. Pseudobulbs are not enough, however, to define most genera in the absence of other characteristics.

Most classification systems have been termed *artificial* instead of *natural,* because they have not always emphasized the evolutionary relationships of the species, genera, sections, subgenera, or other categories in taxonomy. They are practical arrangements, which enable us to communicate information to others. We have had no way of really knowing how *Epidendrum* species actually evolved. We have always assumed that similarity of structure, flower anatomy, presence or absence of certain biochemicals, and, in fact, similar chromosomes or chromosome numbers pointed to close relationships. The use of these characteristics has enabled botanists to produce keys, identify plants, write species or generic descriptions, and in practical terms deal with the wealth of living taxa in the orchid or any other plant family. We have always attempted to make our classification systems as "natural" as possible, trying by deduction and close observation of their comparative anatomy and biochemistry to reflect what *may* have happened in nature. We have never succeeded completely, of course, nor have we ever been able to tell exactly to what degree any system was actually natural rather than artificial. With the orchids, we have also never been able to assess the role of hybridization versus a direct descent through mutation or other factors. A person who deals with these unanswered problems involving evolution may be called a research taxonomist.

The recent use of nucleic acids (DNA), gene analyses, and computers with various systems for cladistic analyses has enabled certain botanists to approach the matter of orchid classification in another fashion. The preliminary data with variously selected compounds of proteins and the nucleic acids has led to modern attempts to classify plants based on analysis by computer correlations. The potential for a more evolutionary basis for classification now becomes possible with these techniques. These classification schemes then lead to the most natural classification, keeping genetic relatives close together and separating those species that are distant genetically, even though alike to our eyes phenotypically. Mark Chase (1986) and others have been publishing bits and pieces of these studies for several years, often producing interesting and unexpected results. Although such systems may have much merit, the results often do not jibe with the more classical systems. With orchids, the genetic studies have only begun to add information that is useful at the generic or species level. The data of DNA analyses compiled have not yet been checked with repeated results by multiple researchers, often do not include data collected from varied analyses of the same species, may not include samples from populations of what are considered type species, and may not be based on the gene-controlled compounds that will produce the best answers. The techniques for obtaining such data are improving all the time, as are the machines, but to write a book about these species *now*, we must mostly content ourselves with the classical data and information on hand, supplementing it when possible with the new data as it becomes available.

We have given the matter much thought and have decided to go ahead with this writing, although we realize the treatment we have proposed may eventually end up completely changed. From our artificial or practical point of view, it is the only avenue open to us until additional species, multiple clones of the same species, and pigments, genes, and proteins that are more varied may be sampled and analyzed. These may be somewhat esoteric scientific thoughts as far as commercial orchid growers, hybridists, or hobbyists may be concerned, but when or if the artificial and the natural systems can be fused, we will likely have a better understanding of the origin, mechanisms, and spread of orchid species, as well as their behavior in nature and in hybrids.

Meanwhile, it would seem to us that paying closer attention to various orchid genera and species that have been proposed in the past, closer adherence to what we do know about the characteristics of those genera and species, and sharpened use of their key differences will add to our present knowledge. We might then be better ready to compare the new classification data based on what may be more objective information provided by biochemistry and computer programs. Until that happy day arrives, we hope that our work may serve a useful function. In the meanwhile, our compilations may prove of help to hobbyists, collectors, and growers by identifying certain epidendroid species, pointing out the use of new or old generic epithets and aiding in the resolution of this interrelated group of species. They are a complex of species, in a manner of speaking, that have been caught or trapped today, as we see them, at one stage in their natural evolution. Let us just hope that the natural environments necessary for orchid evolution are not completely destroyed by mankind to a degree that could eliminate these species in our biosphere, let alone their continued evolution.

Certain people have aided us in this compilation. We would especially like to acknowledge the continuing efforts of Rudolf Jenny, who has patiently been able to supply us with photocopies of many references not readily available through our own somewhat isolated and limited library resources. This, of course, has been a major factor in compiling the necessary data for our research. We happen to live far from most libraries that have suitable runs of older orchid journals in various languages besides English. Rudolf has a wonderful service, Bibliorchidea, which enables researchers in their home to search for orchid-related literature. Without this service, we would have been lost and at the mercy of interlibrary loans.

We are also much indebted to Leslie Garay and to Eric Christenson, who have provided technical advice on many of the taxonomic details that have arisen. We have attempted in all cases to follow the International Code of Botanical Nomenclature, but in many instances, we have found ourselves in botanical muddles that required technical skill beyond our abilities.

Some of the species discussed in this volume have never been illustrated as far as we can tell. They may never have been cultivated as well. In fact, aside from short type descriptions in one old journal or another, some species only exist in our knowledge today because they were collected once. The type specimen may or may not have survived intact in one herbarium or another, and drawings or photos were never made. Until we know there is a holotype specimen available for study today, we can only produce a neotype or lectotype substitute specimen, and there is little else that can be done for the study of some of these poorly known species. Since the time of John Lindley at Kew, no author or botanist has attempted to assemble these debatable species in one place, closing the gaps in the available knowledge about them. We are trying to fill that void with this volume to the extent that is possible.

We hope that the research embodied here may help fill some of these deficiencies, especially if the total record in the world for a given species is limited to the conserved type specimen itself, if it still exists. There is a completely new—or old—research problem here today in presently typifying all of these species with actual herbarium specimens. For most orchid species, the types do still exist, and additional specimens of those species may have found their way to various herbaria of the world. We can hope that our research in preparing this book will ultimately increase the specimens of all these species so that they may all be represented by more than the single specimen of the type example. We are planning to send samples of our specimens to either the Orchid Herbaria of Oakes Ames at Harvard University in Cambridge, Massachusetts (AMES), or to the U.S. National Herbarium at the Smithsonian Institution in Washington, D.C. (US).

Carl is pleased to have his colleague Patricia Harding help with the compilation and writing of this book. Originally majoring in botany before switching to medicine, she has rekindled her interest in plants, raising orchids while fostering the growth and development of her own family. Carl wants to acknowledge that her assistance has given him a large and necessary boost that has made this book possible. As a newly appointed American Orchid Society student judge, she innocently asked him one day if there wasn't some orchid botanical research project she could help accelerate by her input and enthusiasm. Carl knew the answer immediately. What had in part been just an idea for another book now lies before you. Patricia has been particularly good at tracking down the species covered here, many from obscure references, and then converting the type descriptions into a helpful, condensed form. Her keys will be helpful in species identification and the cultural information gathered will help keep these plants in cultivation.

Obtaining photos or drawings of so many poorly known species has plagued us continuously. We have written many letters in our search from Germany to Brazil and México, and around the United States, and other countries too numerous to list, and have been reasonably successful in finding the illustrations necessary for the book. We particularly wish to thank Greg Allikas, Robert Anderson, Noble Bashor, Gernot Bergold, Irene Bock, Dale Borders, Weyman Bussey, Marcos A. Campacci (who contributed pictures, plant descriptions, and some drawings), Eric Christenson, Sandro Cusi, Robert Dressler, Leon Glicenstein, Carlos Hajek, David Hunt, Rudolf Jenny, Roberto Kautsky, Dwayne Lowder, Trudi Marsh, the late Ray McCullough (whose slide legacy is in the hands of Eric Christenson), David Morris, Dan and Marla Nikirk, Ron Parsons, Mauro Peixoto, Andy Phillips (of Andy's Orchids), Art Pinkers, Karlheinz Senghas, Manfred Speckmaier, and Masa Tsubota for their gracious help with the illustrations and for information on cultivation. We thank Wesley Higgins for providing articles and the cladogram used in the text.

We wish to commend the two books, *Botanical Latin* by William Stearn and *Orchid Names and Their Meanings* by Hubert Mayr, whose knowledge, insight, and explanations have helped us throughout the duration of this project. We could not have done it alone.

Jane Herbst, a professionally trained scientific illustrator, did most of the drawings. Jane heard Patricia talking about the book at an orchid society meeting and offered to help. Her skill has brought to our attention many of the attributes the plants had that we weren't aware of and made us rethink several decisions in the keys. She has worked sometimes with real plants but often the drawings were done from less than adequate previous drawings, yet we feel she kept true to what should have been present in the plant characteristics.

Patricia must also thank Mauro Peixoto, who along with his family hosted her in São Paulo, Brazil, for a wonderful tour of orchids in the wild and of local growers. He is an avid orchid grower and just generally nice guy trying to get an orchid tour company going. We recommend him to anyone who would like to see orchids in the wild of Brazil. (For details, see his Web site: www.brazilplants.com.)

Dean Mulyk is to be thanked for helping us understand the DNA cladistic work and its relevance. We thank Marius Wasbauer and Eric Christenson for their help in editing, and Kenton Chambers and Eric Christenson for their help with the Latin work.

Finally, Patricia would like to thank Larry Erickson for giving her Dressler and Pollard's *The Genus* Encyclia *in México*. It was for her the first step of this manuscript.

CARL L. WITHNER, PATRICIA A. HARDING
November 2002

Introduction

"Plant identification is an orderly process resulting in assignment of each individual to a descending series of groups of related plants, as judged by characteristics in common" (Benson 1959). These manmade systems help people deal with the vast variety of plants in the world. The systems are for our use, to help us with identification and to aid in communication. Because these systems depend on human eyes and human criterion for significance, they are imperfect and open to alternative forms. The plants themselves are oblivious to the problems. They have their own pollinators and reproductive mechanisms worked out to their satisfaction.

The study of individual plants and their population groupings is called *classification*, or in more technical jargon, *taxonomy*. A *taxon* is a unit of classification, such as a species, a genus, or a family. Thus taxonomy is the study of taxa.

Botanists, both professional and amateur, spend a lot of time and words manipulating and arranging the species and their variants into systems that represent their views about how the taxa should be arranged. They are constantly looking for discontinuities, as well as similarities, among the populations to arrange the populations into species, the species into genera, and the genera into families and still higher categories in the hierarchy of classification. The variations between individuals in a population provide the means by which new species can arise through mutation, hybridization, and natural and now unnatural selection, especially if a few individuals become isolated from the original population by geographic or other changes that may occur in their habitat. The task of classification is determining the limits when variation within a group exceeds certain parameters and requires making those individuals outside those parameters into a separate group.

Mankind's tendency to group and categorize species into similar clusters or taxa seems an almost inherent characteristic (let's get organized!), and the sharing of certain qualities is generally accepted to show a relationship among the individuals of any given group. However, the arranging of the individuals of a species into larger assemblages becomes more of a problem. There are few objective definitions to guide us in the higher classification categories (as there are, more or less, with species) and this spawns various opinions as to the nature of the larger, more inclusive groups that are being classified. There is no set or objective measure by which botanists can say that a certain group of different species belongs together in one genus, but that a group of different species belongs in another genus, or that still another species belongs in a group by itself.

Botanists search for a way to do this, to find the gaps or discontinuities that distinguish species from species, genera from genera, section from section, but the perfect system for answering the questions has not yet been found. Some researchers depend upon commonalties of the appearance (the *gestalt*) or features of the anatomy; and others examine chromosome numbers, chemical composition of the cells, or microscopic structural details. There is always the hope that one integrated system will finally resolve the problems of classification and at the same time provide a better understanding of how the arrangement ever evolved to match what we are able to observe by studying the individuals. In the case of orchids, the individuals are sometimes variable, and we would like to know the cause of the variability and how to classify them and better understand their interrelationships. What are the essential characteristics that enable us to decide? Is a given plant to be considered as within a given taxon or population, or outside of it?

One fine example of the classification puzzle that is of much interest to many today is what to do with the large and complicated group originally called *Epidendrum*. It was a comparatively simple problem for Linnaeus. He coined the term for orchid plants his students and colleagues were sending back to him from tropical climes, mostly epiphytic orchids that grew upon the branches of trees. Such plants were readily distinguished from the European terrestrial orchids previously known to botanists of that time. The Linnaean term *Epidendrum* has given rise to at least twenty-five to thirty segregate genera. *Epidendrum* still exists as a genus with a residue of species still mixed in that some taxonomists would still prefer to put in separate genera.

Withner discussed this overall group in some detail in *The Cattleyas and Their Relatives*, particularly in Volume IV, where John Lindley's 1853 key (repeated below) to the twelve subgenera of the genus *Epidendrum* is listed with comments. Since Lindley's time some of these twelve subgenera have been raised to generic status, some have been nearly or completely forgotten, and still others have been moved around and incorporated into other genera. For others, still earlier names were found.

In addition, as new orchid species were found, or as other orchid classifiers did their work, sometimes without knowing about Lindley's research in a time with poorer communications and no computers or e-mail, other categories were named and described. These alternative names for *Epidendrum* are listed in the important landmark book written by Oakes Ames, F. Tracy Hubbard, and Charles Schweinfurth working at Harvard University in 1936, *The Genus* Epidendrum *in the United States and Middle America*. The book lists forty synonymous generic epithets for *Epidendrum* and includes most of the twelve subgenera proposed by Lindley in 1853.

Since then, additional generic names have been coined for certain other species segregated as separate taxa because of distinctive characteristics from the large original *Epidendrum* complex. They include among others the segregate genera *Artorima, Dimerandra, Dinema, Epidendropsis, Euchile, Hagsatera, Kalopternix, Nidema,* and so on. Developing a classification is an ongoing process. Needless to say, so is keeping track of all these segregate genera, along with problems about what to do with some of the remaining species still in the genus. We need all the help we can get with these various problems. *Epidendrum* is a puzzle, and the taxonomy of orchids today is anything but a static subject.

Using classical taxonomic techniques one can still employ Lindley's classification proposals for *Epidendrum* as a basis for an overall system of classifying *Epidendrum*, with a synopsis of his subgeneric categories. Lindley's system is remarkable in that it is still useful in many, many aspects of our work, despite the fact that relatively few species had been described and were available to him at the time of his work. He developed and used the classification, and through the various chapters in this book, we will deal with the different sections that he proposed.

One point needs to be emphasized: Each genus, species, or subgroup segregated from *Epidendrum* must be assigned a type species, which is in turn based on a type specimen, an actual single, physical specimen. Ideally, these *holotype specimens,* as they are called, are preserved in one of the major herbaria of the world. There, they may be consulted by orchid specialists interested in determining firsthand the qualities and characteristics upon which our classification systems are based. Should these specimens be destroyed by fire, war, or other catastrophic event, they may be substituted for by duplicate specimens, drawings or other suitable published records that provide what is called a lectotype of the original specimen. If lectotype designation is not possible, a new example, a neotype specimen, can be chosen among other specimens available in the herbaria.

A type specimen must consist of at least a single preserved flower. Lindley's actual type specimens, plants or flowers, are still conserved for study in the Lindley Herbarium at the Royal Botanic Gardens at Kew in Richmond, Surrey, England, and a few other museums such as the Natural History Museum (formerly the British Museum of Natural History). Other recognized herbaria are found throughout the world associated with botanic gardens, universities, or museums and provide the housing for these valuable dried and irreplaceable botanical type specimens. These are the star attractions for scientists, particularly taxonomists, in any such collection. It must be remembered that when taxonomists debate aspects of certain species, one taxonomist will see one trait as more important, while another taxonomist may not even consider this trait. As descriptions of plants are written or read, sometimes traits are not clear; that is why having these plant specimens to provide concrete evidence of "how things really are" is so important.

There are two factions of taxonomists, of course speaking in general terms. Some, the "splitters," separate or split large complex groupings, sections, or genera into smaller groupings. Others, the "lumpers," tend to combine groups into larger groups, doing away with species or genera. The authors of this book think that people can better understand and remember a few entities of closely related plants rather than larger compilations that are more complicated with divergent subdivisions of various degrees.

The purpose of classification is to enable investigation of the natural world and to try to group species together into meaningful clusters. Meaningful, in this case, means by some system that may demonstrate or emphasize how science thinks the various groups or taxa are related and, in turn, the system may show something about the origins and descent of species so that eventually the studies may result in a theoretical family tree (see Dressler 1993) with an evolutionary component.

A species, by most definitions, is a population of individuals that occupy a particular location or habitat, interbreeding together over infinite generations, and thereby perpetuating their traits in nature. They may spread from their original locations into other areas where they may survive and reproduce at random, or they may occasionally mutate and show odd or different characteristics from their parents, thus providing some variation in the population. Plants tend to show a much greater degree of variation than do animals.

Plants of one species may occasionally hybridize with the individuals of another species, producing a yet greater degree of variation in a complex population. Such species complexes resulting from chance hybridization are called *syngameons* in botany. They may become ongoing populations that are composed of parental types of plants along with the hybrids, backcrossings, or sib crossings in all possible chance combinations. Some examples would be the introgressive population of *Cattleya guatemalensis* in Central America, the population of *C. hardyana* that was present in Colombia, or the population of two Brazilian *Cattleya* species and a *Laelia* species that produced the *Laeliacattleya* Elegans population on Santa Catarina Island in Brazil.

Such syngameons arise by the process of introgression, the "invasion" of one species population by the genes of other species. This natural process is not limited to orchid species. It would indicate that genetic incompatibility among the individuals of introgressive species is not as important for keeping them separated as it is in other species. As we researched several of the species for this book, it has become a possibility that some of these syngameous populations may have been placed under a single species epithet, but more than likely, further study on the populations will result in the determination that two or more species are involved.

When a mutant or variable plant or flower turns up in a species population, it is ordinarily designated botanically by the term *forma* in Latin (abbreviated *f.*). These forma or forms are thought to be genetic point or small mutations that occur sporadically. Many color forms, for instance, have turned up in species populations and have been mostly distinguished variously by clonal names. Some have been around for more than a hundred years, propagated by divisions, or more recently by mericloning (the production of new plants from asexual meristem tissue often resulting in thousands of plants, all in theory identical to the parent). These mutations are considered end mutations, ones that do not result in advantageous advancements to reproduction for the plant and the species.

Sometimes one finds differences that appear to be stable components of a population, having either red or white flowers as an example. When one describes this phenomenon then one describes the red variety (var.) of a population or a white variety of a population of one species. The expression of these mutations is thought to be not detrimental to the individual plants but mutations that still allow the plant to propagate itself in nature. Furthermore, this variation is seen repeatedly throughout the population as a certain percentage of the naturally occurring plants.

Sometimes, in nature, a population may be isolated from the main population by strong genetic, ecological, geographic, or geological changes and then will evolve into a population distinct from the main species. When a variant of a species, such as possession of three anthers on the column, occurs in a whole population of plants, usually in a distinct and separate location from the main population of the species, it has been thought of botanically as a subspecies (subsp.). A good example would be the plants of *Anacheilium cochleatum* that produce three-anthered flowers in Florida but not elsewhere in Central America or the Caribbean region where the same species is found. These plants are given the name *A. cochleatum* subsp. *triandrum* to denote that this is a population of plants, distinct from other populations where the three-anthered flowers predominate.

In horticultural circles versus botanical circles, form epithets have traditionally been referred to as varieties without addition of clonal names. This has resulted in confusion when additional plants of the same form have been found and not distinguished except by the varietal name. This double use of the term *variety* in both horticultural and botanical circles causes some confusion for both botanists and growers. When the term *variety* is accompanied by a clonal epithet as well, as should be done, we can usually recognize the difference.

In discussing the individual species in this book, we have not dealt with different forms and have mentioned only limited examples of subspecies of the various species. Given the constraints of pictures and pages for this book, it was more important to stick with the "normal" forms and not branch into the variations. Please keep in mind that a variety you have may be different than what we present here.

Those interested in hybridizing or judging may especially be aware of another problem with the generic names for orchids. The Royal Horticultural Society (RHS) in London keeps the long-term records of hybrids, and for horticultural purposes, it prefers to stay with the large, complex original groupings, such as *Epidendrum*, or *Oncidium*, to mention a second example. Otherwise, the Registrar would forever have to change a variety of hybrid names whenever another segregate genus or hybrid name came along. (At the same time, hobbyists must be wondering why any name changes are necessary at all.) The monumental task of keeping the registry of species and hybrid names has been dutifully and meticulously done by the RHS, and we can only appreciate the tremendous efforts that have gone into its maintenance and accuracy. We know that our past and present knowledge of orchids, as a complex and exciting plant group that cannot be equaled, has probably the best and most complete breeding history of any plant family on record. As far as registering hybrids, despite whatever generic name we give these plants, they will remain epidendrums and encyclias until the RHS system changes its system.

In this age of computers, systems of classification, untainted, one hopes, by the bias of a human classifier, are also being proposed. The morphological matrix aids in producing a cladistic system of classification. This process involves selecting for the plant group in question a number of flower and plant characteristics that can be measured, and giving the characteristics varying weight or significance, or otherwise quantifying or ranking them so that one changes these characteristics into numerical values, which in turn can be fed as data to the computer. The computer then calculates the degree of correlation among and between the characteristics, usually using characteristics of various species, all in theory unprejudiced by human opinion. Depending on the number of species in the sample, the characteristics chosen for correlation in such studies, and the degree to which the correlation must apply, this computed procedure gives rise, then, to a diagram, or cladogram, as it is called.

The clade (a family tree of sorts), it is hoped, provides us with new insight and shows us new relationships among the individuals of a population as they cluster into various groupings. Often the clade or its parts may numerically reconfirm what is already generally accepted to be the relationships in the existing classification of a given group and reinforces our previous opinions. Sometimes it may lead us into new classification schemes. The third potential is that the clade produced provides nothing to add to an existing scheme. The cladograms have the potential to provide new evidence for relationships, but the interpretation of the data must still be provided by the researcher. It would seem from the amount of debate over various schemes that we are not certain we are knowledgeable enough with orchids to turn over our classification problems completely to the computers.

Cladistic systems, incidentally, do not deal readily with natural hybridization, a process that occurs widely in orchids, as a possible factor in species variability. Because most orchid species depend upon and adapt to a specific type of insect pollinator, pollinators can ultimately contribute to species population variability, and therefore the likely evolution of new species from their progenitors, especially within the generic or even the subtribal hierarchies within the orchid family. This, in turn, can complicate the cladistic analyses that depend upon the concept that each present-day species population derives from a similar straight-line ancestral grouping not involving hybridization. It would also seem to imply that any cladistic analysis would begin by analyzing the various species populations involved in any such study to observe what variation already exists among the individuals of those populations. Time, expense, and availability of various clones from particular populations make such studies costly and time consuming. Lack of confidence in trait selection, worth, and assessment have made some present-day researchers skeptical and unwilling to discard the present classification systems based on flower and plant structure, or sometimes, physiological traits. Perhaps the two systems may not be sufficiently compatible in their elementary stages to be used together. Time, repeatability, and acceptance will tell.

This cladistic taxonomy has been greatly aided in recent years by the development of DNA sequencing from the plastid and nuclear genomes of plant material. Using small portions of the genome of a plant, or a component (plastid) contained within the cell containing DNA, a map or sequence of the DNA is obtained that allows an estimation of the total genome, at least to the point of determining relationships. The section of the genome that is chosen for study, in theory mutates from predecessor population's genomes yet should remain constant and fairly uniform within a species and show certain patterns of changes within a particular genus, with even more changes at the section level and so on down the tree to the family level. The study of the genomes of orchid species is being done at a steady rate in recent years, with the results, in bits and pieces, being presented in the literature.

As might be expected in the interesting and far-reaching DNA research, many species populations have yet to have their DNA sampled more than once; and often the type species for some of their sampled genera have not yet been included in the cladograms. Most likely the biggest reason this has not been done is that the procedure is expensive, is dependent on the proper identification of plant material, and the interpretation of the data is a skill one needs to learn requiring dedicated study.

Any correlations between molecularly based data and the classical groupings must be carried out together. In the future when analyses that are more intensive are available, and when there are systems at hand that are better integrated, such correlations may be available. It is obvious already that some molecular data is useful in helping define generic or subgeneric categories, and we look forward to the time when a combination of molecular and classical systems will provide a greater understanding of evolutionary relationships within and among the various orchid subtribes and subfamilies.

Chase (1986), Higgins (1997), and Pridgeon et al. (1999) have provided us with biochemical and cladistic analyses of traits in various orchid groups, and we await their suggestions on amalgamating the classical and the biochemical or cladistic approaches on large and complex groupings such as the subtribes Laeliinae and Oncidiinae. For the moment, however, we seem mostly to be dependent upon the more classically oriented systems and the host of available keys to guide us in the identification of orchid species or genera.

We now go back to Lindley's key to the twelve subgenera of *Epidendrum* that he published in 1853. Most of the key characters are in duplicate (dichotomous), but two are in triplicate (trichotomous). The key is translated into English from Lindley's original Latin. We can only admire the insight that Lindley had into the subdivision of this large, complex group of species as he devised the key. In this book, we are mostly concerned with the species that fall under the category of "lip adnate to the column," the others having been covered in the series *The Cattleyas and Their Relatives*.

Lindley's Key (1853) to the Twelve Subgenera of *Epidendrum*:

Lip nearly free from the column
    Flowers from a spathe . . . . *Epicladium*
    Flowers without a spathe
        Stems pseudobulbous . . . . *Encyclium*
        Stems fusiform . . . . *Diacrium*
Lip adnate to the column
    Stems pseudobulbous
        Flowers without a spathe . . . . *Hormidium*
        Flowers from a spathe, racemose or paniculate

      Flowers arising from the rootstock.................... *Psilanthemum*
      Flowers terminal on the pseudobulb
            Labellum split (lobed) .... *Aulizeum*
            Labellum undivided............................. *Osmophytum*
Stems creeping, with scalelike bracts................................... *Lanium*
Stems slender, leafy erect
    Inflorescence terminal
        Spathe single, large.. *Spathium*
        Spathe several, imbricated........................... *Amphiglottium*
        No spathe .... *Euepidendrum*
    Inflorescence lateral .... *Pleuranthium*

These subdivisions of *Epidendrum* were first published in 1841 in *Hooker's Journal of Botany,* Volume 3, but not arranged as a key. Lindley said there, "Having lately had the occasion to reconsider the large genus *Epidendrum,* I have been led to its subdivision upon more natural characters." Lindley also wrote, "I have had recourse to the organs of vegetation as well as fructification... [T]here seems to be a universal tendency to produce a variety of modifications of the stem and leaves under the same organic type." In other words, the flowers alone were not enough!

A few comments about each of Lindley's subgenera are now appropriate. We already know from the Lindley key what major characteristics separate one subgenus from another. We can also tell from what is known about orchid classification today that many of his subgenera have been changed, some have been dropped or discarded in common usage, though they are still part of the orchid literature; some have been removed because earlier names with priority have been found; several have been elevated to generic status by one investigator or another, even though they may not be commonly used; and, finally, some have been split off into new segregate genera where appropriate. Such splitting is based upon a sharper delineation of what each subgenus or genus should encompass and a better understanding of the species involved in the subgroups. The various taxa in the Lindley key are today mostly considered valid distinctions in a large, varied, and complex group.

We will deal first with the subgenus *Euepidendrum* from Lindley's key. This subgenus, meaning "true" *Epidendrum,* is the group that is still called *Epidendrum* but without the Latin prefix *eu-*. These reed-stemmed plants are common throughout tropical America from México south through the Caribbean and most of South America. They have been the particular interest of Eric Hágsater from México, and he has sorted out many of the confusions and complexities that have beset this large group of species. He has presented many of the results of the research to date in *Icones Orchidacearum* published by the Asociación Mexicana de Orquideologia. Part one of his research, *A Century of New Species in* Epidendrum, was published in 1993. It includes detailed descriptions and very clear and understandable "standardized" drawings of 100 species. *A Second Century of New Species of* Epidendrum was published in 1999, and the *Third Century* has now appeared. We await further publications by Hágsater and his group of researchers and illustrators. With already more than 300 species documented in the orchid literature for this one large taxon alone, one begins to understand not only the degree of complexity of various orchid populations, but also why it is difficult to come up with schemes that orchidologists can agree upon.

Hágsater notes in his 1993 foreword "how poorly the diversity of species is represented in herbaria," and that "we know very little about *Epidendrum*." He went on to state, "It is not until a number of species in a group have been cultivated together, illustrated and studied, that the specific differences become apparent. The distinguishing features vary from one group to another, and what may have been clues to differentiation in one group are totally useless in another." Further, he said, "Unfortunately, traditionally botanists have given the floral segments, the perianth, excessive importance as taxonomic characters, and minimized use of the rest of the plant, including even the other parts of the flower." A key characteristic is one that is unique to the individuals of a given group or species population and therefore can be used to differentiate that population or individual from other members of a family or tribe. The characteristics of a genus are more or less equivalent to a situation where most of the key characteristics are present without major exception. Some are more "critical" than others are, and there is some variability in more minor ways.

Now, turning to more details about the subgenus *Epidendrum* and its characteristics, we list the following points, most of which are key characters in considering the genus. We, in this case, are used to considering or using the epithet both as a subgeneric as well as a generic designation. It must also be used, when necessary, as a sectional epithet. By our international naming rules, the type species of a genus must be in that genus and any subgeneric categories that carry the same epithet as the genus itself. The type species for genus *Epidendrum* is *E. nocturnum* Jacquin. *Epidendrum* is characterized vegetatively by a reedy, non-pseudobulbous leafy stem, which is mostly erect in habit and has a terminal group of flowers without a distinct spathe or sheath at the base of the inflorescence. The individual flower, as Lindley has more or less explained in his texts, shows a lip (lip base or claw) that is adnate or fused by its fleshy base to the edges of the column, the column is hornless and considerably elongated, but not winged, produces four pollinia with caudicles folded back upon themselves, incumbent, and, finally, has a cunicular nectary deeply located between the base of the lip and the base of the column. Pollinating insects have a proboscis long enough to probe the nectary through the elongated narrow channel formed by the callus ridges and the fusion of the column edges with the base of the lip. In so doing, the pollinia of the orchid become attached to the proboscis of the insect when the stylet is withdrawn, and the pollinia are therefore in a position to meet and stick on the stigma when the pollinator probes another flower.

This book is not going to attempt to list or describe any species of the genus *Epidendrum*. Eric Hágsater is currently working on this group with intent to publish, and we await those results. Instead, we will deal with other sections of Lindley's key, focusing on the species with pseudobulbs. This is no small group, but not excessively large either. This makes a nice collection for a book containing interesting and often beautiful species. We are preserving the name *Hormidium*, but will have to replace the names of *Aulizeum* (with *Coilostylis*) and *Osmophytum* (with *Anacheilium*) because the rules of nomenclature require it to be so.

Below is a key to genera covered in the book. Note that *Encyclia* is included—Carl had wanted to have updates to *Encyclia* in the book, but has left that for further publications. We have left the section *Encyclia* in the key and included a brief chapter about it in this book. Furthermore, we are now separating Higgins's large genus *Prosthechea* into other genera (see 2 through 7 below).

Key to the Genera in This Book

1a. Column fused to lip.. *Coilostylis*
1b. Column free of lip .... *Encyclia*
1c. Column fused to lip part way... go to 2
2a. Column teeth equal (except in *Anacheilium radiatum*, which has a fimbriated column midtooth) .... go to 3
2b. Column teeth unequal either in length or in characteristics................ go to 4
3a. Flowers resupinate .... *Pollardia*
3b. Flowers non-resupinate .... *Anacheilium*
4a. Lateral teeth of column longer than medial tooth .... go to 7
4b. Medial tooth of column longer than lateral teeth .... go to 5
5a. Lip longer than 15 mm long... *Panarica*
5b. Lip shorter than 15 mm long.. go to 6
6a. Midtooth of column beaklike (acute) .... *Hormidium*
6b. Midtooth of column rounded, not beaklike .... *Prosthechea*
7a. Lip without warty keels... *Prosthechea*
7b. Lip with warty keels.. *Oestlundia*

It is at this point that we feel we need to address Wesley E. Higgins' work on *Encyclia*, *Prosthechea*, and *Oestlundia*. Most of this work was part of his Ph.D. thesis, completed in 2000, involving both morphological matrixes and DNA plastid data, resulting in several conclusions. The work generated information, which validated some previously published work on the *Epidendrum* subtribe, but also clarified suspicions about several sections of the subtribe that had previously been included in the genus *Encyclia*. For his study he included 30 "in-group taxa" that were to represent all the sections of *Encyclia* based on the work of Dressler and Pollard (1971). He generated a cladogram using the following species listed here as *Encyclia*, but later moved by him into *Prosthechea*: *E. aemula*, *E. chimborazoensis*, *E. cochleata*, *E. cretacea*, *E. fragrans*, *E. glauca*, *E. ionocentra*, *E. ochracea*, *E. prismatocarpa*, *E. pseudopygmaea*, *E. pygmaea*, and *E. vitellina*. (Higgins included many other genera and species in his cladogram; we list only the ones we are concerned with in this volume).

His final cladogram (see page 25) is for weighted holomorphology, where he combines DNA data with morphological parameters of different weighted importance. He generated many trees in his thesis, showing how using different parameters within the tree would change the results, but this last cladogram combines all factors.

His cladogram shows that *Prosthechea aemula*, *P. chimborazoensis*, and *P. fragrans* are closely related; with the later two very close. *Prosthechea cochleata* is slightly more distant from the group but might still be included (in this book we have put these in *Anacheilium*). *Prosthechea ionocentra* and *P. prismatocarpa* are shown closely related to each other but still apart from the *P. fragrans* group (we put them in *Panarica*). *Prosthechea ochracea* is shown more closely related to what we are calling *Anacheilium* and *Panarica* but still outside of those groups. Carl Withner decided the species related to *Encyclia ochracea* belonged in our genus *Prosthechea*, but Patricia Harding suggests that perhaps they need to be in their own group, based on Higgins' data and morphological characteristics. *Prosthechea pygmaea* is separate from the rest (we put them in *Hormidium*) and lastly *Prosthechea cretacea*, *P. glauca*, and *P. vitellina* are remote in their own group (we have these combined with the *Prosthechea ochracea* group to form the genus *Prosthechea*).

Another tree presented in Higgins' thesis, weighted only for combined DNA, shows *Encyclia cochleata* more closely related to *E. pygmaea* than the *E. fragrans* group. *Encyclia glauca* is shown as closely related to *E. ionocentra* and the two of them separated by some distance with the group of *E. cretacea*, *E. ochracea*, and *E. prismatocarpa*. The same tree also shows *Encyclia cyanocolumna*, and *E. tenuissima* being widely separated from *Encyclia luteorosea*, all three of which Higgins later recognized as one genus, *Oestlundia*. It was by bringing in physical characteristics that the human eye sees that he was able to say based on his research the *Oestlundia*

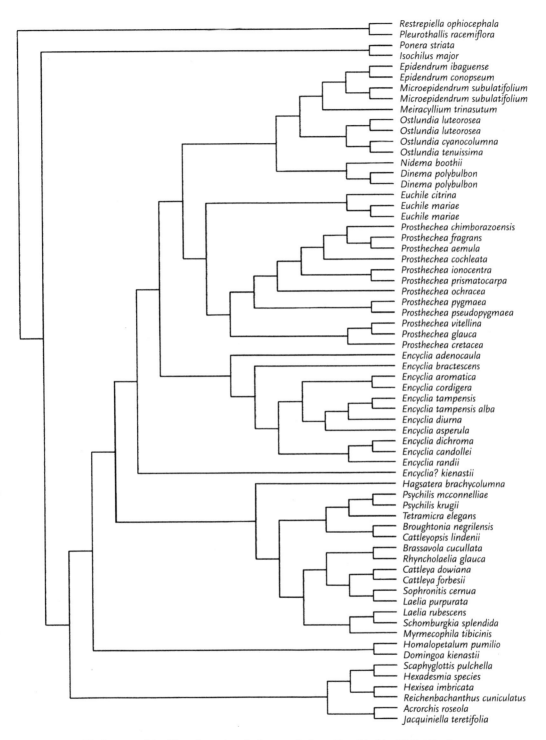

Phylogeny of *Laeliinae* based on holomorphology. Provided by W. E. Higgins.

species are one genus. We do not argue they are one genus; it just seems as if the DNA evidence did not add much information in the *Oestlundia* case.

Now to add to this treasure chest of information, Van den Berg et al. (2000) published another cladogram using only DNA evidence showing *Prosthechea allemanii, P. fausta, P. moojenii, P. suzanense,* and *P. widgrenii* as close relatives, with *P. calamaria* a slightly more distant relative. These species come from Brazil. It also shows the Central American and northern South American species of *P. aemula, P. cochleata,* and *P. venezuelana* as near relatives, but still separated by some distance. Surprising, *P. abbreviata, P. pygmaea,* and *P. vitellina* are closer relatives to the Brazilian group than the Central American species, with *P. abbreviata* being closer to *P. pygmaea* than *P. vitellina*, and all three of them being separate from the previous species grouping. Lying next to all these are *P. lambda, P. linkiana,* and *P. prismatocarpa*. There is no attempt to merge the species of Higgins' project with the species in this project, despite Higgins being one of the coauthors.

If you can follow all that, you must have little idea what these plants look like, because they do not follow what our eyes say should be close relatives. The reassuring part of all this is that it does show trends and we think with time, the testing of more species, and addition of a few more genomes perhaps, a cladogram that does definitely organize this large group of 100 species into genera can be developed. For now, we use what we have. We can certainly see why Higgins left the group as one big genus, and indeed time may show this to be the correct way to look at them, with our divisions being sections of a broadly defined *Prosthechea*. We find, however, that the plants are too diverse to be covered within one genus. We did use the above information in our decisions in dividing up the genera and species, and actually spent much time trying to get physical characteristics to work out with the cladograms, but could not get any firm resolution. We did find it interesting that the Brazilian *Anacheilium* species and the Central American and northern South American *Anacheilium* species were so distantly spaced on the cladogram. We tried to separate the two groups in our key to *Anacheilium* species, but were unable to do so completely.

The only species of our genus *Pollardia* that was tested was *Prosthechea linkiana*. This makes it hard to reach a conclusion on the validity of our grouping of species included within the genus, based on DNA evidence.

CHAPTER 1

The Genus *Anacheilium*

Anacheilium (*Epidendrum* subgenus *Osmophytum* of Lindley) is a group of species most notable for the lip being non-resupinate; the usually unlobed lip turned upwards often in the form of a shell or feather, generally the most attractive portion of the flower. Other orchids in various genera have this trait, but in this large group all the species share the trait. Resupination occurs in most orchid flowers, that is, the flower is actually lip uppermost during development, and, as the bud matures, it rotates on the pedicel to present the lip down. We have all seen the occasional flower that did not turn over correctly, due to environmental conditions. These *Anacheilium* species keep their flowers non-resupinate under all conditions.

Another trait the *Anacheilium* species share is that the inflorescence has a spathe, or covering, that protects the emerging inflorescence. The spathes are either green or green becoming brown, and are sometimes quite large, sometimes rather insignificant. *Encyclia* species do not have this spathe, though *Cattleya* species do, as do many other *Epidendrum* species. The lips of these species often have a cushionlike callus covered with fine hairs like peach fuzz. The species have a three-winged or three-sided fruit (never oval or round). The column rostellum is undivided, and the column has three knobs at the apex (distally), called teeth in the literature. These teeth are equal in length, size, and shape, with a few exceptions; a feature that separates them from other genera in the *Prosthechea* (sense of Higgins) group. Mostly likely, different pollinators determine the need for same versus different shaped teeth. Most *Anacheilium* species are apparently wasp- or bee-pollinated, and many have names based on this relationship (van der Pijl and Dodson 1966).

Another essential characteristic, though not obvious unless the flower or plant is preserved in alcohol, is the presence of unusually numerous fans of raphide (needle-shaped) crystals in the tissues, especially in the flowers. These crystals are apparently glycosidic in composition, but not identical to the oxalate crystal bundles found in many other types of plants, though they may serve the same function. They may be dissolved in hydroxide solutions. They became apparent when plant and flower parts are preserved in alcohol for study, and the resulting dehydration of the tissues made them especially noticeable (see Pabst and Pinto 1981). The raphides, no doubt, provide a deterrent from slugs or other animals, as the needles of the crystals can be painful when penetrating the tongue or linings of the mouth, throat, and digestive tract.

Some researchers consider *Anacheilium* a part of *Encyclia* and others consider it part of *Hormidium*. The key characters of Lindley's subgenus *Anacheilium* must be reviewed more strictly in any scheme for subdividing *Epidendrum*. In preparing this section, we did consider the treatment by Brieger et al. (1977), who subdivided the genus *Osmophytum* into three sections: *Osmophytum*, *Cochleata*, and *Glumacea*. These names are not validly published as sections by just using them in the book, and are of no utility as they were mentioned without descriptions.

Furthermore, it is debated today whether *Osmophytum* or *Anacheilium* should be the correct name for this assemblage, and that requires some research into the history of each name. *Epidendrum cochleatum*, the earliest named species for this section of the cockleshell orchids was given a polynomial by Plumier in 1703 and another by Burmann in 1758 when he published Plumier's work. Linnaeus condensed the polynomials to a binomial epithet in 1763, renaming it *Epidendrum cochleatum*. This species was one of the earliest of New World orchids to be taken to England or Europe for cultivation. Hoffmannsegg in Germany published *Anacheilium cochleatum* in 1842 based on the species *Epidendrum cochleatum*, and thus a new segregate genus from *Epidendrum* was proposed.

Osmophytum as a subgenus was used in Lindley's key in 1853, but he had used the term earlier in 1839 (*Edward's Botanical Register* 29, misc. 135) in describing *Epidendrum inversum* from Brazil, also a cockleshell orchid. Lindley states there, "Of this form of the genus *Epidendrum*, of which *Epidendrum fragrans* may be selected as the type, there are now several species on record, and it is probable that many more remain to be discovered. It will therefore be necessary to provide a distinct section for such species, to which the name *Osmophytum* may be assigned, in allusion to their being usually scented plants." At this point, however, the genus *Anacheilium* would have priority by the rules since *Osmophytum* was only being considered by Lindley as a sectional epithet, not a generic name. We have not found any *Epidendrum* species actually called or published by the generic, not the sectional, epithet of *Osmophytum*, before 1842 when Hoffmannsegg published *Anacheilium cochleatum*, and therefore the latter name has the priority.

The following species are listed by Lindley and others as belonging to this complex, but the list is open to revision as additions, synonymies, and other name changes are worked out. Pabst et al. (1981) have already published many of the species that come from Brazil as *Anacheilium*. In the future, additional species may have to be transferred to the genus.

Dressler (1961) considered these cockleshell species a subgroup of *Encyclia*. He points out several key characteristics: a dorsal, median, erect, large and subquadrate, fleshy, and sometimes fimbriate tooth is separated from a smaller lateral tooth to each side of the column by deep sinuses. In addition, a three-sided seed capsule shape is an indicator of the genus. These traits together provided Dressler with the differences between the two subgroups of *Encyclia*. It seems to us, however, that they just as well provide the basis for a separate genus as for a subgroup. In fact, we still do not quite understand why Dressler considered them a subgroup in *Encyclia*, especially since they form such a well-defined and distinctive group with their non-resupinate flowers and mostly entire midlobes. By including the *Anacheilium* species, the characteristics of *Encyclia* were thereby not so inclusive of all the characteristics, and with our proposed separation, the species within the genus *Encyclia* form a more homogeneous taxon.

The type species for Lindley's *Epidendrum* subgenus *Osmophytum* is *Epidendrum fragrans*, and that epithet would have to be transferred to *Anacheilium*, though it would no longer be the type species of *Anacheilium*, as *A. cochleatum* has priority. Acuña did that in 1938. In the same book Acuña uses *Amphiglottis*, *Aulizeum* (*Auliza*, and so on), *Epicladium*, *Hormidium*, *Nidema*, *Pleuranthium*, and *Seraphyta* as generic names. We would suggest that his research has been ignored until the present, probably because most have not had access to his volume published by the Secretary of Agriculture in Cuba, though there has been a reprint. Higgins (1997) has recently transferred many of these species to the genus *Prosthechea*, and some will have to be corrected, at least for the time being; please see the genus *Prosthechea*, Chapter 8. *Prosthechea* was considered by Lindley to be a subunit of *Encyclia* and was not a taxon under *Osmophytum*.

The name *Anacheilium* means lip turned upwards and this is the most notable feature of these flowers. If this feature were not present in other genera, it would make an easy identifying mark. These plants also have pseudobulbs that arise from a definite rhizome at various lengths, and the pseudobulb has a stalk that holds it apart from the rhizome. The pseudobulbs have sheaths of varying sizes and are ovoid (egg-shaped), elliptic, cylindric, or fusiform (spindle-shaped), and are flattened somewhat, so that there are two sides to the pseudobulb. The column of the flower is attached to the lip for about half of the column length (this fusion varies in percentage). The column is gibbous (flattened on cross section below and rounded above). The rostellum is undivided. The inflorescence has a spathe—a bract that comes out between the uppermost leaves protecting the emerging inflorescence, though at times this is small. The capsule (fruit or seedpod) is three-winged, or three-sided. These plants also have the previously mentioned glycoside flavonoid crystals that become fluorescent under ultraviolet light and should make the flowers more visible in dark areas. This combined with the strong fragrance most likely attracts the insect pollinators.

Culture for this group is mostly warm and wet. Some species tolerate more heat than others do. Consider the origin and especially the elevation as a guide for temperature conditions. Most like moist conditions, high humidity, and little seasonality to the water regime. They like diffuse bright light to full light, but will tolerate medium light intensity. As a group, these produce wonderfully beautiful and fragrant flowers and plants. The flowers are long lived, crystalline in texture, and sparkle in the sun. Plants in cultivation are an attractive green color and keep their leaves several years. They do not sunburn easily nor do they get bacterial spotting, which keeps the plants looking nice.

Many of these species are on any list of plants that are easy to grow in a mixed species collection, being tolerant of various conditions, humidity changes, and over or under watering. They bloom regularly once or twice a year. The blooms last a long time (months) and withstand travel and wind without being bruised like flowers of many other orchid genera. They are also resistant to bugs and slugs—not immune, but not as susceptible as some genera. The flowers have a sparkling crystalline appearance in the sunlight and this sparkle shows up in the photographs quite often. In addition, most are fragrant to some degree or another, some remarkably so. We recommend you try growing a few in your collection, as they are readily available with little searching.

Species of *Anacheilium*

Anacheilium abbreviatum, 40
Anacheilium aemulum, 42
Anacheilium alagoense, 45
Anacheilium allemanii, 46
Anacheilium allemanoides, 48
Anacheilium aloisii, 49
Anacheilium baculus, 49
Anacheilium bennettii, 51
Anacheilium brachychilum, 52
Anacheilium bulbosum, 54
Anacheilium caetense, 55
Anacheilium calamarium, 57
Anacheilium campos-portoi, 59
Anacheilium carrii, 61
Anacheilium chacaoense, 63
Anacheilium chimborazoense, 65

Anacheilium chondylobulbon, 67
Anacheilium cochleatum, 69
Anacheilium confusum, 72
Anacheilium crassilabium, 74
Anacheilium faresianum, 77
Anacheilium farfanii, 78
Anacheilium faustum, 80
Anacheilium fragrans, 81
Anacheilium fuscum, 83
Anacheilium garcianum, 84
Anacheilium gilbertoi, 86
Anacheilium glumaceum, 87
Anacheilium hajekii, 89
Anacheilium hartwegii, 91
Anacheilium ionophlebium, 93
Anacheilium janeirense, 94

Anacheilium jauanum, 96
Anacheilium joaquingarcianum, 98
Anacheilium kautskyi, 98
Anacheilium lambda, 99
Anacheilium lindenii, 101
Anacheilium mejia, 102
Anacheilium moojenii, 104
Anacheilium neurosum, 105
Anacheilium pamplonense, 106
Anacheilium papilio, 107
Anacheilium radiatum, 108
Anacheilium regnellianum, 110
Anacheilium santanderense, 111
Anacheilium sceptrum, 113
Anacheilium sessiliflorum, 115
Anacheilium simum, 116
Anacheilium spondiadum, 118
Anacheilium suzanense, 119
Anacheilium tigrinum, 120
Anacheilium trulla, 122
Anacheilium vagans, 124
Anacheilium vasquezii, 126
Anacheilium venezuelanum, 127
Anacheilium vespa, 128
Anacheilium vinaceum, 130
Anacheilium vita, 131
Anacheilium widgrenii, 133

Key to Species of *Anacheilium*

Before you journey down this key, we wish to explain that for many of these plants we have only seen written descriptions and have had to rely on other authors and what traits they felt were worthy of mentioning. If there are errors, we apologize. In using this key, please note there are two places where you have three or four choices. We have tried to combine the species in groups that we believe to represent relationships, most notably the *Anacheilium faustum-widgrenii* group corresponds to Dressler's *Glumacea* group, and the *Anacheilium aemulum-fragrans* group corresponds to his *Osmophytum* group. It may be that with time, these will be determined to be their own genera, but for now, we include them all in *Anacheilium*. Several species are in two places in the key, this way we felt that if you made a "wrong" turn in the key you might still be able to get to the correct place.

Lip drawings are provided after the key and are divided into four groups correlated to couplets in the key. The drawings may be more useful than the key as many species have a distinctive lip pattern and we were able to get a lip drawing for most of the species. They are not to scale, however, and we apologize for that inadequacy. Your plant may be larger or smaller in cultivation, so the size should only be a relative guide.

1a. Lip concave (bowl-shaped). Column and lip seem to become one basally, junction has margin that wraps around column at least somewhat............ go to 2
1b. Lip reflexed (inverted bowl or dome-shaped) or mostly flat, only concave at edges. Column seems to sit on top of lip, the portion of the lip which is free comes from beneath column, not wrapping or above 180 degrees of horizontal axis of column .... go to 19 a, b, or c

2a. Leaves two or more .... go to 3
2b. Leaf one, lip always pointed... go to 16
3a. Lip very pointed or acute (spear-shaped) .... go to 4
3b. Lip rounded, or if pointed, point is very short .... go to 11
4a. Pseudobulbs clustered or the space between the pseudobulbs less than the height of the pseudobulbs .... go to 5
4b. Plants creeping, with pseudobulbs widely spaced on rhizome (space equal to the length of pseudobulbs) .... go to 9
5a. Flowers always two, back to back on short peduncle.......... *Anacheilium baculus*
5b. Flowers more than two, not back to back .... go to 6
6a. Lip becoming conduplicate (folded) .... go to 7
6b. Lip not folded (conduplicate) .... go to 8
7a. Callus pubescent, tepals without spots .... *Anacheilium lambda*
7b. Callus composed of keels, not pubescent, tepals with spots .... *Anacheilium joaquingarcianum*
8a. Peduncle long (6–13 cm) .... *Anacheilium chondylobulbon*
8b. Peduncle short (10 cm or less) .... *Anacheilium gilbertoi*
9a. Leaves regularly three................................... *Anacheilium vagans*
9b. Leaves usually two... go to 10
10a. Column with pointed wings on lateral edges, lip spade-shaped, flat .... *Anacheilium abbreviatum*
10b. Column without wings, lip cordate at base, forming true heart shape, deeply concave .... *Anacheilium neurosum*
11a. Lip veins red-purple .... go to 12
11b. Lip veins dark maroon, almost black, or brown .... go to 15
12a. Lip notched apically.................................. *Anacheilium ionophlebium*
12b. Lip broad and wide, tapering to a point apically, though may have notch with point sitting within .... go to 13
13a. Lip venation more or less parallel to column............. *Anacheilium chacaoense*
13b. Lip venation radiating out from column, some veins at greater than 90 degrees .... go to 14
14a. Midtooth of column thin, fan-shaped, and toothed, featherlike .... *Anacheilium radiatum*
14b. Midtooth of column thick and fleshy...................... *Anacheilium lambda*
15a. Lip tinged with purple between the lines, reverse (back side-reverse) purple, sepals and petals 13–18 mm .... *Anacheilium trulla*
15b. Lip deep purple almost black, sepals and petals 30–75 mm long................ *Anacheilium cochleatum*
16a. Blooming on immature growth.......................... *Anacheilium aemulum*
16b. Blooming on mature growth .... go to 17

17a. Sepals and petals green to white, lip with purple veins. . . . .
. . . . *A. fragrans* complex, go to 18
17b. Sepals and petals greenish to white with red overlay to almost entirely red with orange tips, lip color cinnamon red . . . . *Anacheilium spondiadum*
18a. Flowers many, lip spade-shaped, tepals basally with a narrow to broadly wedge-like taper, lip 1.7–2.4 cm × 0.9–1.7 cm, column winged. . . . . . *Anacheilium fragrans*
18b. Like *A. fragrans* with smaller flowers, lip measuring only 1.3 cm × 0.8 cm . . . .
. . . . *Anacheilium venezuelanum*
18c. Like *A. fragrans* but with spots at base of the petals . . *Anacheilium chimborazoense*
18d. Like *A. fragrans* but stigma ovate-cordate (rounded heart-shaped), callus slightly depressed, column with deeper clinandrium with shorter lateral teeth . . . .
. . . . *Anacheilium jauanum*
19a. Lip callus squarish like a pad, formed under apex of column and basal but not extending past end of column, not folded or markedly thickened, petals thin. . . . .
. . . . go to 20
19b. Lip callus consisting of keels, sulcate lip not folded, sepals and petals not unusually thick . . . . go to 31
19c. Lip callus folded, thick, sepals and petals thick. . . . . go to 47
20a. Leaf one . . . . *Anacheilium simum*
20b. Leaves two. . . . . go to 21
21a. Pseudobulbs more than 10 cm long . . . . go to 22
21b. Pseudobulbs less than 10 cm long. . . . . go to 25
22a. Flowers green or yellow. . . . . *Anacheilium sceptrum*
22b. Flowers white or pink . . . . go to 23
23a. Flowers mostly white with pink or some combination thereof. . . . . go to 24
23b. Tepals pale pink with blood-red markings. . . . . *Anacheilium allemanoides*
24a. Pseudobulbs 5–11 cm long, more or less cylindrical, flowers white. . . . .
. . . . *Anacheilium faustum*
24b. Pseudobulbs 8–10 cm long, sepals cream colored, lip with short reddish lines at base. . . . . *Anacheilium glumaceum*
24c. Pseudobulbs 15 cm long, sepals white with wine-colored stripes for 2 cm at base (suffused with mauve), lip white, marked with wine stripes . . . .
. . . . *Anacheilium bulbosum*
25a. Sepals and petals solid color without contrasting veins. . . . . go to 26
25b. Sepals and petals with reddish veining or maroon striping. . . . . go to 29
26a. Sepals and petals flesh pink in color (white suffused with green and rose). . . . .
. . . . *Anacheilium allemanii*
26b. Sepals and petals white to green, or yellow. . . . . go to 27
27a. Tepals pure white, flowers 5 cm across . . . . *Anacheilium faustum*
27b. Tepals not pure white, being yellow, flesh pink, or green, flowers 4 cm or less across . . . . go to 28

28a. Flowers 1.5 cm across, yellow on small plant . . . . *Anacheilium sessiliflorum*
28b. Flowers less than 3 cm (but more than 1.5 cm) across, sepals yellow to green. . . . . *Anacheilium calamarium*
28c. Flowers 4 cm across, sepals white to flesh pink colored with green base. . . . . *Anacheilium carrii*
29a. Tepals striped with rose . . . . go to 30
29b. Tepals with dark maroon stripe down center and one stripe on each side of midstripe . . . . *Anacheilium alagoense*
30a. Pseudobulbs 8–10 cm long . . . *Anacheilium glumaceum* (syn. *Epidendrum almasii*)
30b. Pseudobulbs 5–10 cm long . . . . *Anacheilium glumaceum*
31a. Tepals white, green, or yellow . . . . go to 34
31b. Tepals brown-orange, yellowish brown, or flesh pink . . . . go to 32
32a. Flower color flesh pink . . . . *Anacheilium regnellianum*
32b. Flower color other . . . . go to 33
33a. Lip white-yellow to ivory with five pink lines . . . . *Anacheilium campos-portoi*
33b. Lip whitish with nine purple stripes . . . . *Anacheilium kautskyi*
34a. Flower base white . . . . go to 35
34b. Flower base green or yellow. . . . . go to 38
35a. Tepals with red veining . . . . go to 36
35b. Tepals without red veining . . . . go to 40
36a. Lip three-lobed . . . . *Anacheilium garcianum*
36b. Lip one-lobed . . . . go to 37
37a. Flowers many, 12–15, on short 3- to 4-cm-long inflorescence . . . . *Anacheilium caetense*
37b. Flowers few, typically 6, on 15- to 17-cm-long inflorescence . . . . *Anacheilium faresianum*
38a. Sepals and petals 11-nerved. . . . . *Anacheilium moojenii*
38b. Sepals otherwise . . . . go to 39
39a. Petals reflexed. . . . . *Anacheilium widgrenii*
39b. Petals curled forward . . . . *Anacheilium papilio*
40a. Lip unlobed, oblong, and pointed . . . . *Anacheilium suzanense*
40b. Lip lobed, though perhaps obscure, not oblong, but about as long as wide . . . . go to 41
41a. Sepals keeled. . . . . *Anacheilium aloisii*
41b. Sepals unkeeled . . . . go to 42
42a. Lip stiff, not bendable. . . . . go to 43
42b. Lip flexible in texture. . . . . go to 46
43a. Lip pointed. . . . . go to 44
43b. Lip rounded or notched . . . . go to 45
44a. Petals with spots or bars arranged transversely . . . . *Anacheilium crassilabium*
44b. Petals with longitudinal stripes . . . . *Anacheilium vespa*

45a. Lip rounded	*Anacheilium tigrinum*
45b. Lip notched, squarish	*Anacheilium pamplonense*
46a. Lip yellow with few purple flecks	*Anacheilium santanderense*
46b. Lip with maroon venation	*Anacheilium mejia*
46c. Lip with solid dark maroon center	*Anacheilium vita*
47a. Midlobe of lip much larger than medial lobes	go to 48
47b. Medial and lateral lobes similar in size	go to 52
48a. Flower pink to red	*Anacheilium garcianum*
48b. Flower yellow, green, or brown, often with spots	go to 49
49a. Lip pointed	go to 50
49b. Lip rounded or notched	go to 51
50a. Petals with spots or bars arranged transversely	*Anacheilium crassilabium*
50b. Petals with longitudinal stripes	*Anacheilium vespa*
51a. Lip rounded	*Anacheilium tigrinum*
51b. Lip notched, squarish	*Anacheilium pamplonense*
52a. Callus extends onto midlobe of lip	go to 56
52b. Callus does not extend onto midlobe of lip	go to 53
53a. Lip base color dark (red, purple, or brown)	go to 54
53b. Lip base color light (yellow, green, or white)	go to 57
54a. Sepals brown-orange	go to 55
54b. Sepals other colored	go to 61
55a. Sepals brown-orange, petals dark brownish red, lip purple with white margin	*Anacheilium bennettii*
55b. Sepals and petals dark brown with yellow border, lip yellowish with few purple spots	*Anacheilium janeirense*
56a. Lip white with lavender suffusion	*Anacheilium hajekii*
56b. Lip yellow with splash of red	*Anacheilium brachychilum*
57a. Sepals solid color	*Anacheilium lindenii*
57b. Sepals with blotches, spots, or mottling	go to 58
58a. Column white	go to 59
58b. Column purple	*Anacheilium farfanii*
59a. Lip white with purple marks	*Anacheilium hartwegii*
59b. Lip red or white with pink suffusion	go to 60
60a. Lip dark red with white margin	*Anacheilium vasquezii*
60b. Lip white suffused with pink and purple, finely streaked with purple	*Anacheilium* undescribed Peruvian plant
61a. Sepals and petals brown with lavender suffusion	*Anacheilium fuscum*
61b. Sepals and petals deep purple, lip cerise-purple	*Anacheilium vinaceum*

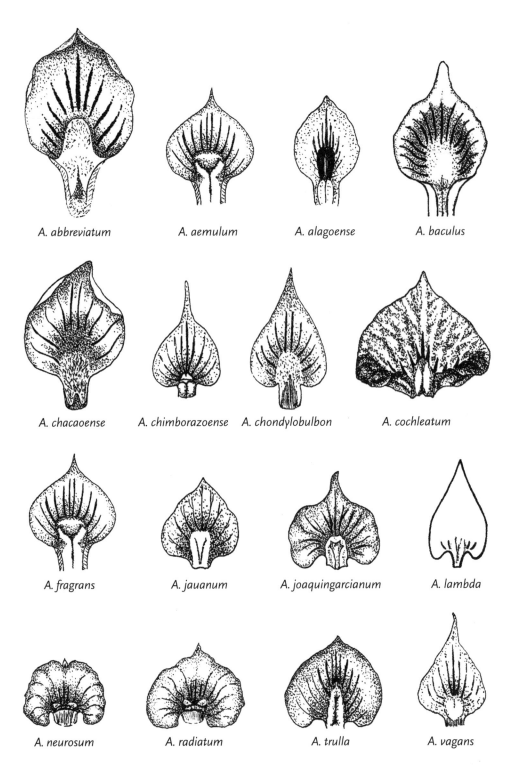

Figure 1-1. *Anacheilium* lips, corresponding to couplets 3a–18c in the key to species of *Anacheilium*.

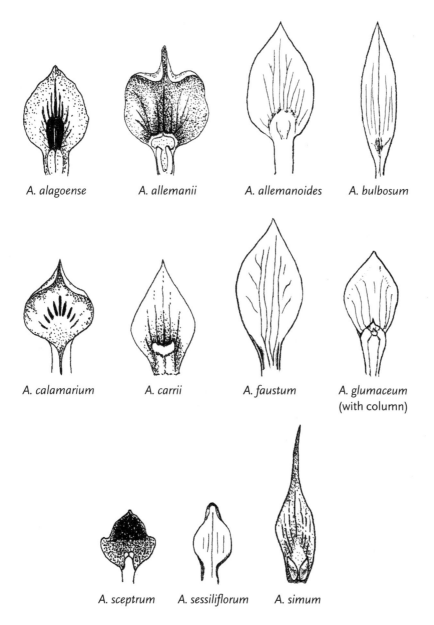

Figure 1-2. *Anacheilium* lips, corresponding to couplets 19a–30b in the key to species of *Anacheilium*.

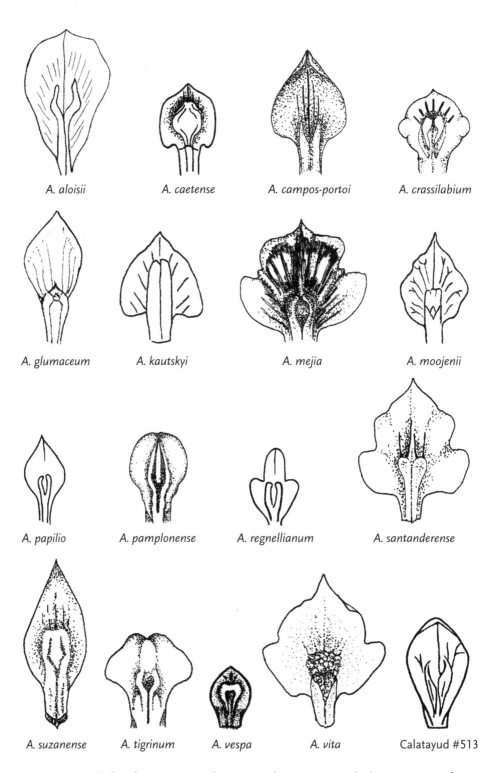

Figure 1-3. *Anacheilium* lips, corresponding to couplets 31a–46c in the key to species of *Anacheilium*.

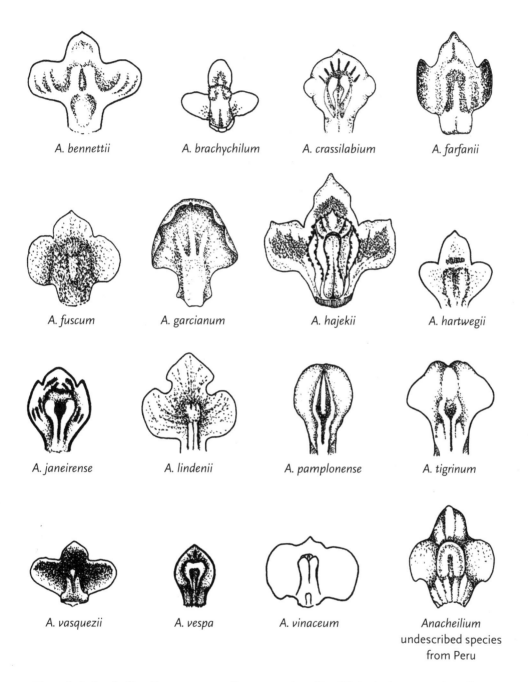

Figure 1-4. *Anacheilium* lips, corresponding to couplets 47a–61b in the key to species of *Anacheilium*.

Anacheilium abbreviatum
FIGURE 1-5, PLATE 1

Anacheilium abbreviatum (Schlechter) Withner & Harding, *comb. nov.* Basionym: *Epidendrum abbreviatum* Schlechter. 1906. *Repertorium Specierum Novarum Regni Vegetabilis* 3: 107. Type: Costa Rica: without number or location, 10 May 1913, *Tonduz 17618* (holotype: B, destroyed; isotype: F).

SYNONYMS
Epidendrum prorepens Ames. 1923. *Schedulae Orchidianae* 2: 33.
Encyclia abbreviata (Schlechter) Dressler. 1961. *Brittonia* 13: 264.
Prosthechea abbreviata (Schlechter) W. E. Higgins. 1997. *Phytologia* 82 (5): 381.

DERIVATION OF NAME
Latin *abbreviatus*, "abbreviated," referring to the short inflorescence.

DESCRIPTION
An epiphyte 10–20 cm tall. Rhizome creeping, elongated between pseudobulbs, and rooting at the intervals. Pseudobulbs stalked (stipitate), narrow widening out to bulb and narrowing again, flattened, glossy when young. Leaves two. Flowers stalk short, nearly sessile; though can be up to 4 cm long, with sheath at base. Flowers one to six, non-resupinate, greenish white, with linear purple spots on tepals, lip marked with purple. Callus yellow. Lip ovate, acute, strongly concave, white with few longitudinal purple stripes, fleshy, margin slightly undulate. Column with triangular erect lobule on each side at the summit. Capsule three-angled.

HABITAT AND DISTRIBUTION
Costa Rica, Ecuador, Guatemala, Honduras, México, Nicaragua, Panama, and Peru. In wet montane forest, at 100–1400 meters in elevation.

FLOWERING TIME
May to August.

CULTURE
Culture for this species is warm to intermediate temperatures in wet conditions, meaning frequent watering with a fair-draining media that retains some moisture. These plants should be able to withstand not being watered for some time but will do better when given constant moisture, especially in the spring and summer months.

COMMENT

Three plants—*Anacheilium abbreviatum* (Plate 1), *A. alagoense* (Plate 3), and *A. vagans* (Plate 45)—appear very similar, having the red line down the tepals. Compare the description carefully to make certain of your identification. As the species are separated by wide geographic distances, it is unlikely they are close relatives, and more likely they are examples of convergent evolution.

This species is a rambler with long distances between the pseudobulbs, which can make it difficult to maintain in collections due to space considerations.

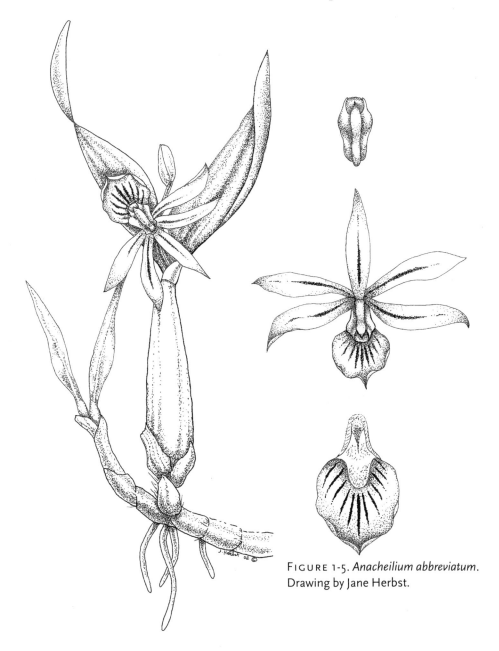

FIGURE 1-5. *Anacheilium abbreviatum*. Drawing by Jane Herbst.

MEASUREMENTS
Pseudobulbs 3–7.5 cm long, 0.4–1.2 cm wide
Leaves 7–16 cm long, 0.4–0.9 cm wide
Inflorescence 1.5 cm long
Spathe 4 cm long
Sepals 13–15 mm long, 2.5–3 mm wide
Petals 12–12.5 mm long, 2.2.5 mm wide
Lip 8.5–10 mm long, 5–7.5 mm wide
Column 5–5.5 mm long

Anacheilium aemulum
FIGURE 1-6, PLATE 2

Anacheilium aemulum (Lindley) Withner & Harding, *comb. nov.* Basionym: *Epidendrum aemulum* Lindley. 1836. *Botanical Register* 22: t. 1898.

SYNONYMS
Epidendrum cordatum Vellozo. 1827. *Florae Fluminensis* 9: t. 38.
Encyclia fragrans var. *aemula* (Lindley) Reichenbach f. 1853. *Linnaea* 25: 243.
Epidendrum aemulum var. *brevistriatum* Reichenbach f. 1876. *Linnaea* 41: 37.
Epidendrum apiculatum Lindley. 1877. *Linnaea* 41: 81.
Epidendrum fragrans var. *aemulum* (Lindley) Barbosa Rodrigues. 1881. *Gen. Sp. Orch. Nov.* 2: 136.
Epidendrum fragrans var. *janeirense* Barbosa Rodrigues. 1881. *Gen. Sp. Orch. Nov.* 2: 137.
Epidendrum fragrans var. *rivularium* Barbosa Rodrigues. 1881. *Gen. Sp. Orch. Nov.* 2: 137.
Epidendrum fragrans var. *micranthum* Barbosa Rodrigues. 1881. *Gen. Sp. Orch. Nov.* 2: 138.
Epidendrum fragrans var. *alticallum* Barbosa Rodrigues. 1881. *Gen. Sp. Orch. Nov.* 2: 138.
Epidendrum fragrans var. *brevistriatum* (Reichenbach f.) Cogniaux. 1898. *Flora Brasiliensis (Martius)* 3 (5): 85.
Encyclia fragrans subsp. *aemula* (Lindley) Dressler. 1971. *Phytologia* 21: 440.
Encyclia aemula (Lindley) Carnevali & Ramírez. 1993. *Monographs in Systematic Botany* 45: 1257.
Prosthechea aemula (Lindley) W. E. Higgins. 1997. *Phytologia* 82(5): 381.

DERIVATION OF NAME

Latin *aemulans,* "rivaling, more or less equaling," referring to the plant's similarity to *Anacheilium fragrans.*

DESCRIPTION

An epiphyte. Rhizome stout and creeping. Leaf one. Inflorescence of three to five flowers produced on emerging new growth. Flowers non-resupinate, greenish

FIGURE 1-6. *Anacheilium aemulum.*
Drawing by Jane Herbst.

white with 15 parallel purple stripes on lip. Sepals lanceolate, acute recurved, petals oblong, acuminate (pointed), narrowed at base. Lip simple, concave, ovate, acuminate, narrowed at base, callus three-lobed, basal, conforming to column. Column broad, apically three dentate, with straplike tip to the column. Flowers fragrant.

HABITAT AND DISTRIBUTION

Brazil, Ecuador, Guyana, French Guiana, Panama, Peru, Surinam, Trinidad and Tobago, Venezuela, and actually most of South America. In wet montane forest at 650–1800 meters in elevation.

FLOWERING TIME

July to December.

CULTURE

Culture for these plants is intermediate to warm temperatures, watering to maintain damp media throughout the year, with daily watering in the spring and summer. They grow in nature in areas of high humidity and almost daily rain during most of the year. They will tolerate long periods (weeks) of drying out, but would prefer to be watered frequently. They can be mounted or potted, making sure that if mounted they are watered frequently. Light requirements are bright filtered sunlight.

COMMENT

This flower strongly resembles the flower of *Anacheilium fragrans*. *Anacheilium aemulum* has traditionally been kept as a separate species from *A. fragrans* based on the feature of blooming on the immature growth. We have decided to leave it be, though we think you could argue, perhaps correctly, that they are the same species. According to the *Botanical Register*, the difference between *A. fragrans* and *A. aemulum* is that *A. aemulum* has pseudobulbs that are exactly ovoid, not tapered, smaller flowers, and sepals and petals of same length. Both species are widespread geographically, found in warm wet forests.

There is quite a bit of variation in flower and plant type and actually, there is probably some work that needs to be done defining this species, as Brazil has several varieties that do not fit within the above description.

MEASUREMENTS

Pseudobulbs 5 cm long, 2.5 cm wide
Leaves 42 cm long, 2.5 cm wide
Inflorescence 5 cm long
Sepals 25 mm long, 6 mm wide
Petals 23 mm long, 11 mm wide
Lip 16 mm long, 12 mm wide
Column 8 mm long, 3.8 mm wide

Anacheilium alagoense
PLATE 3

Anacheilium alagoense (Pabst) Pabst, Moutinho & A. V. Pinto. 1981. *Bradea* 3: 23. Basionym: *Epidendrum alagoense* Pabst. 1963. *An. 14th Cong. Soc. Bot. Bras.*: 18. Type: Brazil, Alagoas, Sítio do Espéto, da Usina Serra Grande, Municipality of São José da Laje ad extremum finem Munic. Pernambucense "Canhotinho" (10.12.61.) *Dr. Luiz de Araujo Pereira s.n.* (holotype: HB 19741).

SYNONYMS
Encyclia alagoensis (Pabst) Pabst. 1967. *Orquídea* 29 (6): 276.
Prosthechea alagoensis (Pabst) W. E. Higgins. 1997. *Phytologia* 82 (5): 381.

DESCRIPTION
Pseudobulbs spindle-shaped, slightly compressed. Leaves two. Flowers one or two, on short raceme. Sepals fleshy, yellow green, dorsal sepal seven-nerved, lateral sepals nine-nerved. Petals yellow-green, five-nerved, with a dark maroon stripe down the midvein, and one other dark stripe lateral to the midvein. Lip fleshy, ovate, bluntly pointed, attached to the column partway, lip with parallel maroon markings. Callus basally oblong, apex shallowly notched.

HABITAT AND DISTRIBUTION
Brazil (Alagoas, Pernambusco).

COMMENT
Three plants—*Anacheilium abbreviatum* (Plate 1), *A. alagoense* (Plate 3), and *A. vagans* (Plate 45)—appear very similar, having the red line down the tepals. Compare the description carefully to make certain of your identification.

MEASUREMENTS
Pseudobulbs 4–6 cm long, 0.5–0.8 cm wide
Leaves 7–11 cm long, 0.6–0.8 cm wide
Inflorescence 1–2 cm long
Spathe 1 cm long
Sepals 11–12 mm long, 3–3.5 mm wide
Petals 11 mm long, 3 mm wide
Lip 11 mm long, 7 mm wide

Anacheilium allemanii
FIGURE 1-7, PLATES 4, 5

Anacheilium allemanii (Barbosa Rodrigues) Pabst, Moutinho & A. V. Pinto. 1981. *Bradea* 3: 23. Basionym: *Epidendrum allemanii* Barbosa Rodrigues. 1877. *Gen. Spec. Orch. Nov.* 1: 54. Type: Brazil, not preserved.

SYNONYMS
Encyclia allemanii (Barbosa Rodrigues) Pabst. 1972. *Orquídea* 29 (6): 276.
Hormidium allemanii (Barbosa Rodrigues) Brieger. 1977. Schlechter's *Die Orchideen*, ed. 3, p. 571.
Prosthechea allemanii (Barbosa Rodrigues) W. E. Higgins. 1997. *Phytologia* 82 (5): 381.

DESCRIPTION
Pseudobulbs short, narrowly ovoid-ellipsoid. Leaves two. Flowers four to six. Sepals and petals flesh pink in color (white suffused with green and rose). Lip unlobed, white with pink vein extending from callus (sometimes with dark purple spots). Column short, thick necked with club shape, central tooth blunt and broader than lateral teeth. Ovary three-winged.

Plate 4 shows a plant awarded in the United States by the American Orchid Society and Plate 5 a plant in Brazil, which we assume was photographed in its natural environment. We do not know if the difference in the red color between the two is cultural or genetic.

HABITAT AND DISTRIBUTION
Brazil (Minas Gerais). Epiphytic in tall trees at 1200–1400 meters in elevation.

FLOWERING TIME
August to September.

CULTURE
Culture for this plant is warm to intermediate temperatures, daily water in spring and summer, less water in fall and winter, high humidity, and diffuse bright light.

MEASUREMENTS
Pseudobulbs 7 cm long, 3 cm wide
Leaves 20 cm long, 4 cm wide
Inflorescence 18 cm long
Spathe 8 cm long

Sepals 25 mm long, 8 mm wide
Petals 23 mm long, 9 mm wide
Lip 18 mm long, 18 mm wide
Column 1 mm long

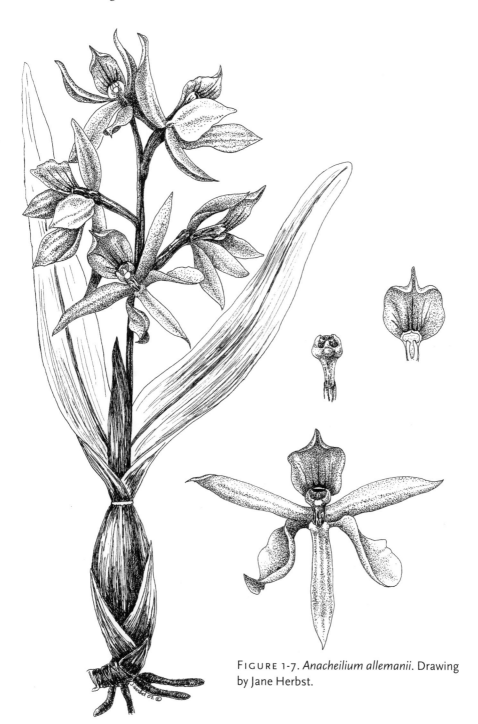

FIGURE 1-7. *Anacheilium allemanii*. Drawing by Jane Herbst.

Anacheilium allemanoides
PLATE 6

Anacheilium allemanoides (Hoehne) Pabst, Moutinho & A. V. Pinto. 1981. *Bradea* 3: 23. Basionym: *Epidendrum allemanoides* Hoehne. 1933. *Bol. Agr. São Paulo* 616. t. 8. Type: Brazil (holotype: SP #26.680).

SYNONYMS
Encyclia allemanoides (Hoehne) Pabst. 1967. *Orquídea* 29 (6): 276.
Hormidium allemanoides (Hoehne) Brieger. 1977. Schlechter's *Die Orchideen*, ed. 3, p. 571.
Prosthechea allemanoides (Hoehne) W. E. Higgins. 1997. *Phytologia* 82 (5): 381.

DESCRIPTION
An erect epiphyte, with a short, thick, creeping rhizome. Pseudobulbs sheathed, spaced 1.2 to 2 cm apart on the rhizome. Leaves two or three, intense green. Inflorescence densely multiflowered. Flowers non-resupinate, pale pink with blood-red markings, petals twisted. Lip entire, ovate, adnate to column basally, on the inside marked blood red, edge entirely white. Column short, apex thickened, held high, white with blood-red lines.

HABITAT AND DISTRIBUTION
Brazil (Espírito Santo, Minas Gerais, and São Paulo).

COMMENTS
This species was first thought to be *Anacheilium bulbosum* but the lip is shaped differently. There is also a possibility that it is a natural hybrid, per the describing authors: *Epidendrum* ×*allemanoides* Hoehne. 1934. *Bol. Agr. São Paulo* 34: 16, which is *Epidendrum allemanii* × *E. inversum* (synonym *E. bulbosum*).

MEASUREMENTS
Pseudobulbs 7–10 cm long, 2.5–2 cm wide
Leaves 15–25 cm long, 2–3 cm wide
Inflorescence 7–11 cm long
Spathe 1.5 cm long
Sepals 22–25 mm long, 5 mm wide
Petals 7–8 mm wide (length not given)
Lip 10 mm long, 10 mm wide
Column 8–9 mm wide

Anacheilium aloisii

Anacheilium aloisii (Schlechter) Withner & Harding, *comb. nov.* Basionym: *Epidendrum aloisii* Schlechter. 1921. *Repertorium Specierum Novarum Regni Vegetabilis* 8: 66. Type: Ecuador, Imbabura, epiphytic in the mountains of Mojanda, near San José de Abinas, July 1871, *L. Sodiro 54* (holotype: B, destroyed; lectotype: QPLS designated by Dodson in Jorgensen and Leon 1998).

SYNONYM
Prosthechea aloisii (Schlechter) Dodson & Hágsater. 1999. *Monographs in Systematic Botany* 75: 956.

DESCRIPTION
A robust epiphyte. Pseudobulbs cylindric, slightly compressed. Leaves two or three. Flowers five to eight, flesh-pink colored, yellow, brown blotched, sepals keeled, sepals slightly smaller than petals. Lip entire, clawed, short, narrow, finely pointed, sinus on edge of lip fringed by short hairs. Callus horseshoe-shaped of two parallel keels, each keel ending in a raised area. Column three-toothed, fleshy, lateral tooth extends more. Capsule three-angled.

HABITAT AND DISTRIBUTION
Ecuador.

MEASUREMENTS
Pseudobulbs 18–20 cm long, 1 cm wide
Leaves 17–22 cm long, 3–4 cm wide
Inflorescence 18 cm long
Spathe 4–5 cm long
Sepals 18 mm long
Petals smaller than sepals
Lip 13 mm long, 7.5 mm wide
Column 7 mm long

Anacheilium baculus
PLATE 7

Anacheilium baculus (Reichenbach f.) Withner & Harding, *comb. nov.* Basionym: *Epidendrum baculus* Reichenbach f. 1856. *Bonplandia* 4: 214.

SYNONYMS

Epidendrum pentotis Reichenbach f. 1876. *Linnaea* 41: 81.
Epidendrum acuminatum Sessé & Mociño. 1894. *Fl. Mex.* ed. 2: 202.
Epidendrum beyrodtianum Schlechter. 1915. *Orchis* 9: 49, t. 4, f. 14–21.
Encyclia pentotis (Reichenbach f.) Dressler & Pollard. 1961. *Brittonia* 13: 265.
Encyclia baculus (Reichenbach f.) Dressler & Pollard. 1971. *Phytologia* 21 (7): 436.
Hormidium baculus (Reichenbach f.) Brieger. 1977. Schlechter's *Die Orchideen*, ed. 3, p. 569.
Prosthechea baculus (Reichenbach f.) W. E. Higgins. 1997. *Phytologia* 82 (5): 381.

DERIVATION OF NAME

Latin *baculus,* "stick" or "rod," referring to the pseudobulbs.

DESCRIPTION

An epiphyte. Pseudobulbs clustered, large, spindle-shaped, cylindrical, elongate. Leaves two. Flowers two, large for genus (7 cm across), held well visible from foliage with back of lips touching. Sepals and petals green-yellow to cream-colored. Lip concave, inverted, heart-shaped, yellowish to whitish with purple nerves radiating outward on disc of lip. Callus oblong, thickened sulcate in the center at the base, to occasionally two distinct short parallel keels. Ovary roughly warty, three-winged. Flowers very fragrant.

HABITAT AND DISTRIBUTION

Belize, Brazil, Colombia, Costa Rica, El Salvador, Guatemala, Honduras, México, and Nicaragua. Found at 400–1700 meters in elevation.

FLOWERING TIME

April to June.

CULTURE

This plant is easy to grow and bloom, tolerant of a wide range of conditions and neglect. We grow ours in the Pacific Northwest in baskets, hanging high among the vandas getting nearly full northwest sun (which is not as bright as California or Florida). The plants are watered every day in the spring and summer, weekly in the fall and winter, with high humidity in the spring and summer and lower in the winter. These plants grow in a clump and can remain undisturbed in a container for several years. In cultivation, we have seen plants that get large and others that seem to stay small, with flowers of all sizes.

MEASUREMENTS

Pseudobulbs 10–20 cm long, 1–1.5 cm wide
Leaves 20 cm long, 2–2.5 cm wide
Inflorescence 2.5–3.5 cm long

Spathe 4 cm long
Sepals and petals 40 mm long, 6–7 mm wide
Column 8 mm long, 3 mm wide

Anacheilium bennettii
FIGURE 1-8, PLATE 8

Anacheilium bennettii (Christenson) Withner & Harding, *comb. nov.* Basionym: *Encyclia bennettii* Christenson. 1994. *Brittonia* 46 (1): 29. Type: Peru. Dept. Huanuco: Prov. Huanuco, near Carpish Pass on Huanuco side, 2350 m, 31 August 1985, *D. & A. Bennett 3490* (holotype: USA).

SYNONYM
Prosthechea bennettii (Christenson) W. E. Higgins. 1997. *Phytologia* 82 (5): 381.

DERIVATION OF NAME
Honors David E. Bennett, Jr., an expert on Peruvian orchids and coauthor of the illustrated series *Icones Orchidacearum Peruvianum*.

DESCRIPTION
A terrestrial on rocks, or an epiphyte to 40 cm tall. Rhizome woody, creeping, enveloped by short scarious bracts. Pseudobulbs cylindric, lightly compressed, slender, enveloped by loosely sheathing bracts to beyond the middle, comprising one long internode and one short apical internode (between the leaves). Leaves two or three. Flowers spreading, sepals brownish orange, petals dark brownish red, lip purple with white margin. Sepals subequal. Lip three-lobed with central fleshy quadrangular callus, lateral lobes orbicular, midlobe quadrangular with acute triangular apex. Column stout, tridentate (three-toothed).

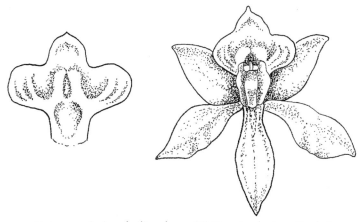

FIGURE 1-8. *Anacheilium bennettii*. Drawing by Jane Herbst.

HABITAT AND DISTRIBUTION
Peru. Found at 2000–4150 meters in a climate of reduced rainfall, among shrubs, herbaceous plants, and grass in clay loam soil with organic litter.

FLOWERING TIME
August to October.

CULTURE
We do not know cultural requirements for this plant. Using the elevation of the plant's habitat as a guide, we would guess at cool to intermediate temperatures. We would grow it in a plastic pot, watering more in the spring and summer, but this is only an educated guess.

MEASUREMENTS
Pseudobulbs 12–17 cm long, 1 cm wide
Leaves 20 cm long, 1.75 cm wide
Inflorescence 18 cm long
Spathe 7 cm long
Sepals 11 mm long, 4.5–5 mm wide
Petals 9 mm long, 5 mm wide
Lip 3.5 mm long, 3.7 mm wide

Anacheilium brachychilum
FIGURE 1-9, PLATES 9, 10

Anacheilium brachychilum (Lindley) Withner & Harding, *comb. nov.* Basionym: *Epidendrum brachychilum* Lindley. 1846. *Orch. Linden.* 9.

SYNONYMS
Epidendrum pachyanthum Schlechter. 1919. *Repertorium Specierum Novarum Regni Vegetabilis, Beihefte* 6: 38.
Encyclia brachychila (Lindley) Carnevali & Ramírez. 1986. *Ernestia* 36: 9.
Prosthechea brachychila (Lindley) W. E. Higgins. 1997. *Phytologia* 82 (5): 381.

DERIVATION OF NAME
Latin *brachy,* "short," and *chilus,* "lipped," referring to the lip.

DESCRIPTION
An epiphyte with a tough woody rhizome, without sheaths. Pseudobulbs lightly compressed, lower half clothed in sheaths. Leaves two or three. Flowers up to 10. Sepals and petals yellow spotted with brown. Lip fleshy, lateral lobes yellow cream,

rather variable shape, midlobes with splash of light maroon-red. Callus yellow cream with fine hairs. Anthers fleshy yellow. Flowers very fragrant.

HABITAT AND DISTRIBUTION

Colombia, Venezuela, and possibly Ecuador. In dense forests on the slopes of the Sierra Nevada at 2000 meters in elevation.

FLOWERING TIME

August.

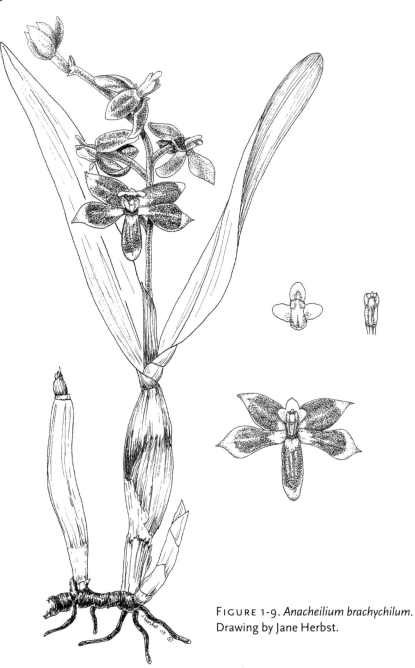

FIGURE 1-9. *Anacheilium brachychilum*. Drawing by Jane Herbst.

CULTURE

Culture for this plant is intermediate to cool temperatures, constant moisture in spring and summer with less frequent watering in fall and winter. We grow ours in a hanging basket where it is watered every day in the growing season.

COMMENTS

Plates 9 and 10 show two plants growing in Ecuador at Ecuagenera, the nursery of Pepe Portillo and his family. They demonstrate the variation of color.

MEASUREMENTS

Pseudobulbs 10 cm long
Leaves 25 cm long, 2.5 cm wide
Inflorescence 17 cm long
Sepals 15 mm long, 6 mm wide
Petals 13 mm long, 7 mm wide
Lip 10 mm ling, 10 mm wide
Column 10 mm long

Anacheilium bulbosum
PLATE 11

Anacheilium bulbosum (Vellozo) Withner & Harding, *comb. nov.* Basionym: *Epidendrum bulbosum* Vellozo. 1831. *Florae Fluminensis* 9: t. 11.

SYNONYMS

Epidendrum inversum Lindley. 1839. *Edward's Botanical Register* 25, misc. 85.
Epidendrum latro Reichenbach f. ex Cogniaux. 1898. *Flora Brasiliensis (Martius)* 3 (5): 82.
Encyclia bulbosa (Vellozo) Pabst. 1967. *Orquídea* 29 (6): 276.
Encyclia inversa (Lindley) Pabst. 1976. *Bradea* 2 (14): 81.
Anacheilium inversum (Lindley) Pabst, Moutinho & A. V. Pinto. 1981. *Bradea* 3: 23.
Prosthechea inversa (Lindley) W. E. Higgins. 1997. *Phytologia* 82 (5): 381.
Prosthechea bulbosa (Vellozo) W. E. Higgins. 1997. *Phytologia* 82 (5): 381.

DERIVATION OF NAME

Latin *bulb*, "bulb," and *osum*, "full," for the pseudobulb. This plant was first described as an *Epidendrum* and since it had a pseudobulb, the epithet distinguished it from other *Epidendrum* species.

DESCRIPTION

Pseudobulbs elongate, ovoid, compressed, 2 cm apart on horizontal rhizome. Leaves two or three, strap-shaped. Flowers six to fifteen. Sepals and petals same shape and size though subequal. Sepals white with wine-colored stripes for 2 cm at base (suffused with mauve). Lip pointed, folded back on itself basally, lip tip with three small calli. Lip white marked with wine stripes. Column three-toothed, obtuse. Column wine-striped. Ovary three-winged.

HABITAT AND DISTRIBUTION

Brazil (Espírito Santo, Guanabara, Minas Gerais, Paraná, Rio de Janeiro, Rio Grande do Sul, São Paulo, Santa Catarina) and Paraguay. Found in the mid to upper tree zone or on rocks in cloud forest, in filtered light with high humidity, at 1000–1300 meters in elevation.

FLOWERING TIME

January to May.

MEASUREMENTS

Pseudobulbs 15 cm long, 1.8 cm wide
Leaves 22 cm long, 3 cm wide
Inflorescence 20 cm long
Spathe 7 cm long
Sepals 20–22 mm long, 2–2.5 mm wide
Petals 18–22 mm long, 3–4 mm wide
Lip 16–18 mm long, 4–6 mm wide
Column 7 mm long

Anacheilium caetense

FIGURE 1-10, PLATE 12

Anacheilium caetense (Bicalho) Pabst, Moutinho & A. V. Pinto. 1981. *Bradea* 3: 23. Basionym: *Hormidium caetense* Bicalho. 1973. *Bol. Soc. Camineira Orq.* 2 (4): 26. Type: Brazil, Minas Gerais, Caeta (holotype: SP 118045).

SYNONYMS

Encyclia caetensis (Bicalho) Pabst ex The International Plant Name Index, as *Encyclia caetense*.
Prosthechea caetensis (Bicalho) W. E. Higgins. 1997. *Phytologia* 82 (5): 381.

DERIVATION OF NAME
After Caeta, a city in Brazil, and Latin *ensis*, "place," referring to the location it was originally found.

DESCRIPTION
An epiphyte, sometimes a lithophyte, up to 40 cm tall. Pseudobulbs suborbicular, slightly compressed. Leaves two or three. Flowers 12–18, fleshy, about 3 cm in diameter, pale yellow with dark red veins, petals slightly paler. Lip unlobed, squarish, apex slightly pointed, pale yellow with fine red linear markings and network of red blotches near margin of callus. Callus squarish, depressed. Column basally red-striped, anther yellow.

HABITAT AND DISTRIBUTION
Brazil (Bahia and Minas Gerais).

COMMENT
The photograph of this plant and the drawing published with the type description remind us of the *Anacheilium vespa* type of plants. We have never seen this plant, only the picture of it, so have no idea how much substance the flower parts have, but if they were somewhat thick we would wonder if this species shouldn't be considered in the *Anacheilium vespa* complex, to be studied later.

MEASUREMENTS
Pseudobulbs 9–12 cm long, 3–4 cm wide
Leaves 22–27 cm long, 3–4 cm wide
Inflorescence 12 cm long
Spathe 3–3.5 cm long
Sepals 15–18 mm long, 6.5–7.8 mm wide
Petals 13–16 mm long, 6–8 mm wide
Lip 10 mm long, 10 mm wide

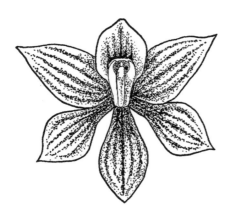 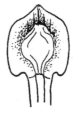

FIGURE 1-10. *Anacheilium caetense*. Drawing by Jane Herbst.

Anacheilium calamarium
FIGURE 1-11, PLATE 13

Anacheilium calamarium (Lindley) Pabst, Moutinho & A. V. Pinto. 1981. *Bradea* 3 (23): 183. Basionym: *Epidendrum calamarium* Lindley. 1838. *Edward's Botanical Register* 24, misc. 88.

SYNONYMS

Epidendrum pipio Reichenbach f. 1856. *Allgemeine Gartenzeitung* 24: 98.
Epidendrum punctiferum Reichenbach f. 1881. *Gardener's Chronicle Agricultural Gazette* 2: 38.
Epidendrum calamarium var. *longifolium* Cogniaux. 1898. *Flora Brasiliensis (Martius)* 3 (5): 89.
Epidendrum calamarium var. *brevifolium* Cogniaux. 1898. *Flora Brasiliensis (Martius)* 3 (5): 89.
Epidendrum calamarium var. *latifolium* Cogniaux. 1898. *Flora Brasiliensis (Martius)* 3 (5): 89.
Epidendrum organense Rolfe. 1898. *Kew Bulletin* 194.
Hormidium calamarium (Lindley) Brieger. 1961. *Publicacao Cientifica Universidade de São Paulo, Institut de Genetica* 2: 69.
Encyclia calamaria (Lindley) Pabst. 1967. *Orquídea* 29 (6): 276.
Encyclia organensis (Rolfe) Pabst. 1967. *Orquídea* 29 (6): 276.
Encyclia pipio (Reichenbach f.) Pabst. 1967. *Orquídea* 29 (6): 277.
Encyclia punctifera (Reichenbach f.) Pabst. 1967. *Orquídea* 29 (6): 277.
Anacheilium punctiferum (Reichenbach f.) F. Barros. 1983. *Hoehnea* 10: 85.
Prosthechea calamaria (Lindley) W. E. Higgins. 1997. *Phytologia* 82 (5): 381.
Prosthechea pipio (Reichenbach f.) W. E. Higgins. 1997. *Phytologia* 82 (5): 381.
Prosthechea punctifera (Reichenbach f.) W. E. Higgins. 1997. *Phytologia* 82 (5): 381.

DERIVATION OF NAME

Greek *calos*, "beautiful," and Latin *marinus*, "growing in the sea." Calamari is a small squid and the name refers to the fanciful resemblance of the flower to squid.

DESCRIPTION

A small mat-forming epiphyte. Pseudobulbs, clustered, spindle-shaped, slightly compressed, arising from rhizome at 1-cm intervals, light green. Leaves two or three. Roots tend to lift plant above substrate. Flowers four to six on dense raceme, non-resupinate, color white to slightly green, with five small narrow violet (near blue) streaks on lip parallel to column. Column three-toothed, lateral teeth longer and acute. Ovary three-winged.

HABITAT AND DISTRIBUTION

Brazil (Espírito Santo, Guanabara, Minas Gerais, Paraná (*punctiferum*), Rio de Janeiro, São Paulo) and Venezuela. Found at 1400 meters (220–600 meters another source) on thick branches or main trunk of trees.

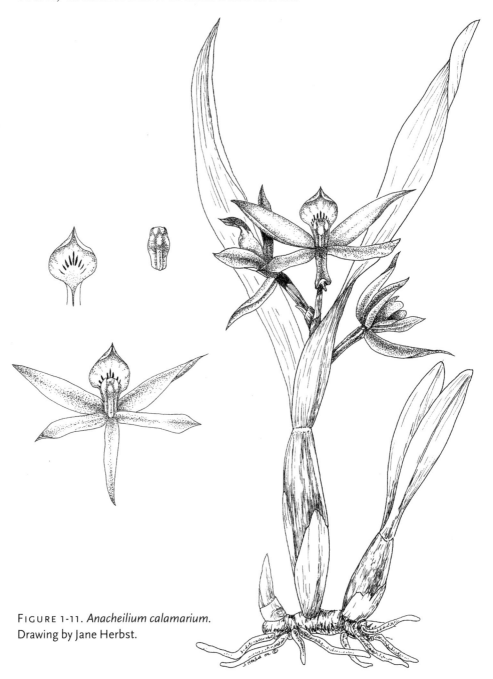

FIGURE 1-11. *Anacheilium calamarium*.
Drawing by Jane Herbst.

FLOWERING TIME
December and April to May.

CULTURE
These plants need intermediate to warm conditions with frequent watering to do well, though will survive in other conditions. They do well in pots, baskets, or mounted. Humidity should be high, less in winter. Fertilize throughout the year. Mauro Peixoto, who lives and grows orchids in Brazil, says he only fertilizes his plants twice a year.

COMMENT
Dunsterville reported no callus on the lip, only thickening on basal part; Christenson confirms that the callus is quite low.

We decided to combine *Epidendrum punctiferum* and *E. pipio* under this species; however, these names are still sometimes used. Both of these variations have lips that are more trowel-shaped and pointed. *Epidendrum punctiferum* has a raceme that is 3–8 cm long, with flowers that are larger than the typical *Anacheilium calamarium*. *Epidendrum pipio* has a raceme that bears only two or three flowers.

MEASUREMENTS
Pseudobulbs 3–5 cm long, 0.6–0.8 cm wide
Leaves 5–7 cm long, 0.6–0.9 cm wide
Inflorescence 2–4 cm long
Spathe 4–6 mm long,
Sepals 9–10 mm long, 1.5–2 mm wide
Petals 8–9 mm long, 1.5 mm wide
Lip 7 mm long, 5–6 mm wide
Column 5 mm long

Anacheilium campos-portoi
FIGURE 1-12, PLATE 14

Anacheilium campos-portoi (Pabst) Pabst, Moutinho & A. V. Pinto. 1981. *Bradea* 3: 23. Basionym: *Encyclia campos-portoi* Pabst. 1967. *Orquídea (México)* 29: 62, tab 2. Type: Brazil, Espírito Santo, near Domingos Martins, 580 meters, flowered in cultivation 3 February 1964, *R. Kautsky 67* (holotype: HB 20.970).

SYNONYM
Prosthechea campos-portoi (Pabst) W. E. Higgins. 1997. *Phytologia* 82 (5): 381.

DERIVATION OF NAME
Honors Paulo de Campos Pôrto, former director of the Rio de Janeiro Botanic Gardens.

DESCRIPTION
A small epiphyte with a thick rhizome bent in zigzag fashion, ash gray in color. Pseudobulbs slightly wrinkled, compressed with sheath that covers much of the

FIGURE 1-12. *Anacheilium campos-portoi*. Drawing by Jane Herbst.

pseudobulb. Leaf one. Flowers three or four, arising from bract, fleshy but firm. Perianth pale brown, lip white-yellow to ivory with five pink lines. Disk of lip has five keels. Lip pointed, column flesh-pink colored.

HABITAT AND DISTRIBUTION
Brazil (Espírito Santo).

FLOWERING TIME
January to February in Brazil.

CULTURE
Patricia saw this plant in bloom in January in Mauro Peixoto's greenhouse near São Paulo, Brazil. Mauro was growing it in a plastic web pot in sphagnum moss mixed with tree fern fiber. He waters it frequently and fertilizes it twice a year.

COMMENT
This plant is very often mislabeled as *Anacheilium kautskyi*, and indeed several pictures were submitted as being one or the other. You can identify these two species from a photograph by the leaf count: *A. campos-portoi* has one leaf, and *A. kautskyi* has two rounded leaves. Vitorino Paiva Castro Neto, Roberto Kautsky, and Marcos Campacci have helped us determine the photographs reproduced here.

MEASUREMENTS (based on the type illustration)
Pseudobulbs 1–5.2 cm long, 1.2–1.5 cm wide
Leaves 4.5–5 cm long, 1.8–2 cm wide
Inflorescence 5 cm long
Spathe 2.5 cm long
Sepals 12 mm long, 2–2.5 mm wide
Petals 9 mm long, 2 mm wide
Lip 8 mm long, 4 mm wide
Column 5 mm long

Anacheilium carrii
FIGURE 1-13, PLATE 15

Anacheilium carrii (Neto & Campacci) Withner & Harding, *comb. nov.* Basionym: *Prosthechea carrii* Neto & Campacci. 2001. *Orchid Digest* 65 (2): 74.

DERIVATION OF NAME
Honors George F. Carr, a specialist in *Cycnoches* and related genera.

DESCRIPTION

A lithophyte. Pseudobulbs obovate (not exceeding twice as long as broad), elongate, slightly flattened. Leaves two or three. Inflorescence with a spathe that covers the peduncle completely. Flowers five. Sepals and petals white with greenish base, lip white with few purple veins near the base, callus white. Lip entire, basally subrectangular, apically sharply triangular. Column three-toothed, medial tooth shaped like a fist.

HABITAT AND DISTRIBUTION

Brazil (Bahia). Found at 1200–1600 meters in elevation.

FLOWERING TIME

March.

MEASUREMENTS

Pseudobulbs 3.4–4 cm long, 2–2.5 cm wide
Leaves 9 cm long, 2.5 cm wide
Inflorescence 5 cm long
Sepals 19–20 mm long, 4 mm wide
Petals 18 mm long, 4–4.5 mm wide
Lip 12–13 mm long, 5–6 mm wide
Column 6 mm long, 3 mm wide

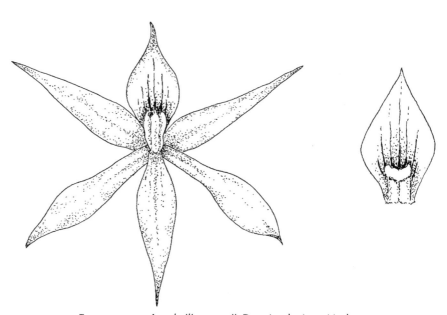

FIGURE 1-13. *Anacheilium carrii*. Drawing by Jane Herbst.

Anacheilium chacaoense
FIGURE 1-14, PLATE 16

Anacheilium chacaoense (Reichenbach f.) Withner & Harding, *comb. nov.* Basionym: *Epidendrum chacaoense* Reichenbach f. 1854. *Bonplandia* 2: 20.

FIGURE 1-14. *Anacheilium chacaoense*. Drawing by Jane Herbst.

SYNONYMS

Epidendrum pachycarpum Schlechter. 1906. *Repertorium Specierum Novarum Regni Vegetabilis* 3: 109.

Encyclia chacaoensis (Reichenbach f.) Dressler & Pollard. 1971. *Phytologia* 21: 436.

Hormidium chacaoense (Reichenbach f.) Brieger. 1977. Schlechter's *Die Orchideen*, ed. 3, p. 569.

Prosthechea chacaoensis (Reichenbach f.) W. E. Higgins. 1997. *Phytologia* 82 (5): 381.

DERIVATION OF NAME
After Chacao, a region in South America.

DESCRIPTION
Pseudobulbs clustered, ovoid or ellipsoid, compressed. Leaves two. Inflorescence with six to eight non-resupinate flowers. Flowers 4–5 cm in diameter. Sepals and petals greenish white. Lip widens then tapers to a point apically, unlobed, deeply cupped. Lip has five straight purple lines radiating from base but mostly parallel to column. Callus pubescent and centrally grooved. Capsule three-angled.

HABITAT AND DISTRIBUTION
México to Panama, Colombia, and Venezuela. Found at 250–1200 meters in open deciduous oak forest, sometimes on rocks.

FLOWERING TIME
March to July.

CULTURE
This species comes from warm temperature zones with fairly constant moisture year round. We grow this plant mounted, with frequent watering year-round.

COMMENT
This species is considered variable, and a lot of the variability may just be confusion with *Anacheilium ionophlebium*. The latter can be distinguished by a notch on the lip apex and a lip that is not fleshy thickened, usually marked with purple lines. Some authors consider these two species to be synonyms.

MEASUREMENTS
Pseudobulbs 4–10 cm long, 2–5 cm wide
Leaves 15–35 cm long, 1.7–5 cm wide
Inflorescence 4–10 cm long
Spathe 3 cm long

Sepals 14–35 mm long, 5–10 mm wide
Petals 12–24 mm long, 6–10 mm wide
Lip 10–20 mm long, 8–25 mm wide
Column 10–12 mm long

Anacheilium chimborazoense
FIGURE 1-15, PLATE 17

Anacheilium chimborazoense (Schlechter) Withner & Harding, *comb. nov.* Basionym: *Epidendrum chimborazoense* Schlechter. 1916. *Repertorium Specierum Novarum Regni Vegetabilis* 14: 389.

SYNONYMS
Encyclia chimborazoensis (Schlechter) Dressler. 1971. *Phytologia* 21: 440.
Hormidium chimorazoenze (Schlechter) Brieger. 1977. Schlechter's *Die Orchideen*, ed. 3, p. 568.
Prosthechea chimborazoensis (Schlechter) W. E. Higgins. 1997. *Phytologia* 82 (5): 381.

DERIVATION OF NAME
After Chimborazo Province, Ecuador, one of the places where the species grows.

DESCRIPTION
An epiphyte. Pseudobulb oblong, compressed. Leaf one, lanceolate, acute. Growths are floppy, tending to grow horizontally, not erect. Flowers 3–7 (up to 15), large, up to 5 cm in diameter, non-resupinate. Sepals and petals white sparsely speckled toward the base with dark purple. Lip white striped with reddish purple lines, pointed at apex, cupped. Ovary three-winged. Strongly fragrant of overbearing honey.

HABITAT AND DISTRIBUTION
Ecuador to Panama, and northern South America. Found in tall trees and in trees overhanging rivers.

FLOWERING TIME
Most of the year, particularly February to August.

CULTURE
Culture is like that of most others in this group: intermediate conditions, frequent watering, and high humidity.

COMMENT

This species is distinguished from *Anacheilium fragrans* by spots at base of the petals and by petals with consistently revolute (pinched) lateral margins toward the base. It is pollinated by wasps.

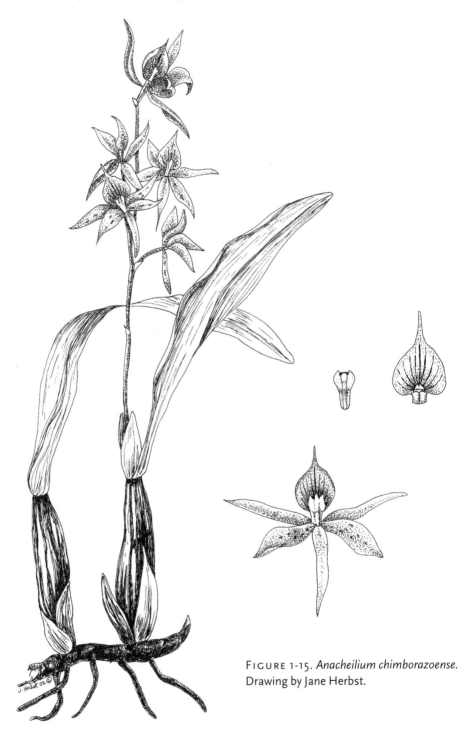

FIGURE 1-15. *Anacheilium chimborazoense*. Drawing by Jane Herbst.

MEASUREMENTS

Pseudobulbs 10 cm long, 2 cm wide
Leaves 17 cm long, 2.4 cm wide
Inflorescence 12 cm long
Spathe 4 cm long
Sepals 29 mm long, 6 mm wide
Petals 23 mm long, 8 mm wide
Lip 20 mm long, 13 mm wide
Column 6 mm long

Anacheilium chondylobulbon
FIGURE 1-16, PLATE 18

Anacheilium chondylobulbon (A. Richard & Galeotti) Withner & Harding, *comb. nov.* Basionym: *Epidendrum chondylobulbon* A. Richard & Galeotti. 1845. *Annales des Sciences Naturelles Botanique*, sér. 3, 3: 20.

SYNONYMS

Encyclia chondylobulbon (A. Richard & Galeotti) Dressler & G. E. Pollard. 1971. *Phytologia* 21 (7): 436.
Hormidium chondylobulbon (A. Richard & Galeotti) Brieger. 1977. Schlechter's *Die Orchideen*, ed. 3, p. 569.
Prosthechea chondylobulbon (A. Richard & Galeotti) W. E. Higgins. 1997. *Phytologia* 82 (5): 381.

DERIVATION OF NAME

Latin *chondyl*, "cartilage," and *bulbo*, "bulb," referring to the pseudobulb.

DESCRIPTION

Plant to 50 cm tall. Rhizome thick and abbreviated. Pseudobulb spindle-shaped, flattened slightly. Leaves three to five. Flowers five. Sepals and petals lanceolate, lip unlobed, abruptly long, acuminate (pointed), concave below with lateral margins upcurved. Petals and sepals greenish yellow to cream-colored, lip whitish with purple nerves. Column light green with short brown purple basal lines. Rectangular callus at the lip base is shallowly grooved. Capsule three-winged.

HABITAT AND DISTRIBUTION

El Salvador, Guatemala, México, and Nicaragua. Found in large masses on rocks in partial shade in humid forests at 1000–2600 meters in elevation.

FLOWERING TIME
June to September.

CULTURE
Culture is the same as others in this group: intermediate temperatures with frequent watering in spring and summer, and less in fall and winter.

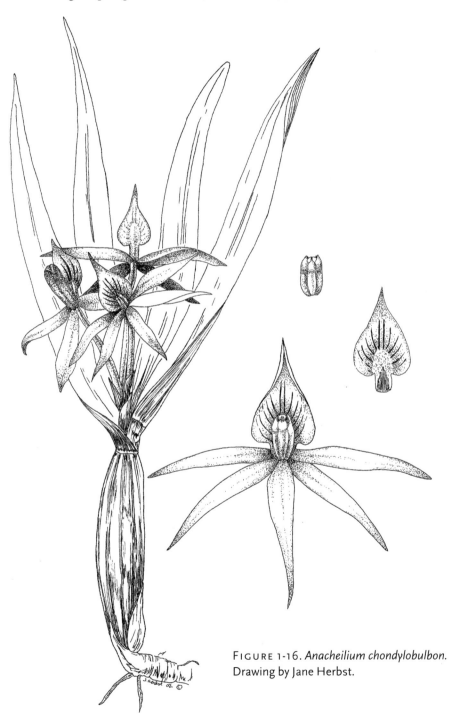

FIGURE 1-16. *Anacheilium chondylobulbon.* Drawing by Jane Herbst.

COMMENT

Species is closely allied to *Anacheilium radiatum*, but *A. chondylobulbon* usually has more leaves and a shorter isthmus (claw) connecting the lip fan to the column, contrasting to *A. radiatum* with never more than three leaves and with a distinct isthmus at the base of the lip before widening out to a transverse blade. This species is also similar to *A. chacaoense* but *A. chondylobulbon* has pseudobulbs that are more elongated, leaves more numerous, and flowers that are larger with longer, more pointed lips.

MEASUREMENTS

Pseudobulbs 7–25 cm long, 1.2–2.2 cm wide
Leaves 3–14 cm long, 1–2.2 cm wide
Inflorescence 6–13 cm long
Sepals 25–37 mm long, 3–6.5 mm wide
Petals 22–26 mm long, 4–5.5 mm wide
Lip 17–22 mm long, 8–11 mm wide
Column 5–6 mm long

Anacheilium cochleatum
FIGURE 1-17, PLATE 19

Anacheilium cochleatum (Linnaeus) Hoffmannsegg. 1842. *Litteratur-Bericht zur Linnaea (Berlin)* 16: 229. *Verzeichniss der Orchideen* 21. Basionym: *Epidendrum cochleatum* Linnaeus. 1763. *Sp. Pl.*, ed. 2, 1351.

SYNONYMS

Epidendrum lancifolium Pavón ex Lindley. 1831. *Gen. Sp. Orch. Pl.* 98.
Epidendrum cochleatum var. *pallidum* Lindley. 1853. *Fol. Orch. Epid.* 41.
Aulizeum cochleatum (Linnaeus) Lindley ex Stein. 1892. *Orchideenbuch* 227.
Phaedrosanthus cochleatus (Linnaeus) O. Kuntze. 1903. *Post. Ktze. Lex. Gen. Phanerog.* 429.
Epidendrum cochleatum var. *costaricense* Schlechter. 1923. *Repertorium Specierum Novarum Regni Vegetabilis, Beihefte* 19: 118.
Encyclia cochleata (Linnaeus) Lemée. 1955. *Flore de la Guyane Française* 1: 418; isonym: *Encyclia cochleata* (Linnaeus) Dressler. 1961. *Brittonia* 13: 264.
Encyclia lancifolia (Lindley) Dressler & Pollard. 1971. *Phytologia* 21: 437.
Hormidium cochleatum (Lindley) Brieger. 1977. Schlechter's *Die Orchideen*, ed. 3, p. 569.
Prosthechea cochleata (Linnaeus) W. E. Higgins. 1997. *Phytologia* 82 (5): 381.

DERIVATION OF NAME
Latin *cochleatus*, "spoon-shaped," referring to the deeply concave lip.

COMMON NAMES
Octopus orchid, cockleshell orchid.

DESCRIPTION
A stout plant. Pseudobulbs stalked (stipitate), ovoid or ellipsoid, strongly compressed. Leaves one to three, elliptic, acute. Inflorescence loosely flowered on raceme or few-branched panicle, to 45 cm long. Ovary 1–4 cm long. Sepals and

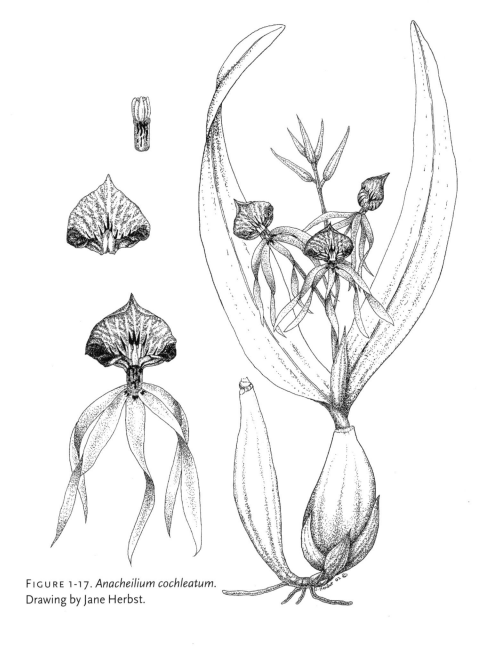

Figure 1-17. *Anacheilium cochleatum*.
Drawing by Jane Herbst.

petals greenish white or greenish yellow with purple blotches at bases, helically twisted and strongly reflexed. Lip spreading from the middle of the column, deep purple with basal central portion whitish, with conspicuous radiating purple veins, broadly cochleate (shape similar to a clam shell), deeply concave, abruptly pointed at apex, with somewhat undulate margins. Callus of two or three basal keels or swellings. Capsule three-winged.

HABITAT AND DISTRIBUTION
México to Panama throughout the West Indies and northern South America. Found from near sea level to 1900 meters, in tropical evergreen, deciduous, and oak forests.

FLOWERING TIME
Almost everblooming.

CULTURE
Ideal conditions would be warm temperatures, high humidity, and frequent watering.

COMMENT
Anacheilium cochleatum is a plant for everyone's collection, tolerant of all sorts of conditions and abuse, be it low or high temperatures, drought or over watering, bugs, and slugs. In addition, it rewards you with a spike that, though long, commonly produces new blooms over 18 months. By the time that spike is done, the new growth is well in bloom, so established plants are virtually everblooming.

MEASUREMENTS
Pseudobulbs 5.5–26 cm long, 2–5 cm wide
Leaves 20–33 cm long, 3–5 cm wide
Inflorescence 50 cm long
Spathe 2–11 cm long
Sepals 30–75 mm long, 3–7 mm wide
Petals 30–75 mm long, 3–7 mm wide
Lip 10–21 mm long, 13–26 mm wide
Column 7–9 mm long

Anacheilium cochleatum var. *triandrum*

Anacheilium cochleatum var. ***triandrum*** (Ames) Small. 1933. *Man. Southeast. Fl.* 392. Isonym: *A. cochleatum* var. *triandrum* (Ames) Sauleda, Wunderlin & Hansen. 1984. *Phytologia* 56 (4): 308. Basionym: *Epidendrum cochleatum* var.

triandrum Ames. 1904. *Contr. Orch. Fl. S. Fla.* 16. Holotype: USA, Florida, illustration by Blanche Ames, plate VIII, published with the original description.

SYNONYMS
Epidendrum triandrum (Ames) House. 1906. *Muhlenbergia* 1: 129.
Encyclia cochleata subsp. *triandra* (Ames) Hágsater. 1993. *Orquídea (México)* 13 (1–2): 215.
Encyclia cochleata var. *triandra* f. *albidoflava* P. M. Brown. 1995. *North American Native Orchid Journal* 1 (2): 131.
Prosthechea cochleata var. *triandra* (Ames) W. E. Higgins. 1999. *North American Native Orchid Journal* 5: 18.
Prosthechea cochleata var. *triandra* f. *albidoflava* (P. M. Brown) P. M. Brown. 1999. *North American Native Orchid Journal* 5: 18.

DERIVATION OF NAME
Latin *triandrus,* "three-anthered," for the flowers.

DESCRIPTION
This variety form has three anthers instead of the usual two and is usually a less robust plant with smaller flowers.

HABITAT AND DISTRIBUTION
South Florida, Dominican Republic, and Puerto Rico.

Anacheilium confusum
FIGURE 1-18

Anacheilium confusum (Rolfe) Withner & Harding, *comb. nov.* Basionym: *Epidendrum confusum* Rolfe. 1899. *Orchid Rev.* 7: 197.

SYNONYM
Epidendrum fragrans var. *megalanthum* Lindley. 1849. *Jour. Hort. Soc. London* 4: 223.

DESCRIPTION
Pseudobulbs tall. Leaves two. Flowers 10 cm in diameter, on a relatively short inflorescence. Lip larger than that of *Anacheilium baculus* and less pointed. Flowers straw colored with numerous radiating red-purple lines of the lip, whereas flowers of *A. baculus* seem to have whiter sepals and petals and the lip is heavier purple-shaded.

HABITAT

Guatemala.

MEASUREMENTS

Pseudobulbs 30 cm long, 2.5 cm wide
Leaves 30 cm long, 3 cm wide
Inflorescence 7 cm long
Sepals and petals 50 mm long, 15 mm wide
Lip 25 mm long, 15 mm wide
Column 10–15 mm long

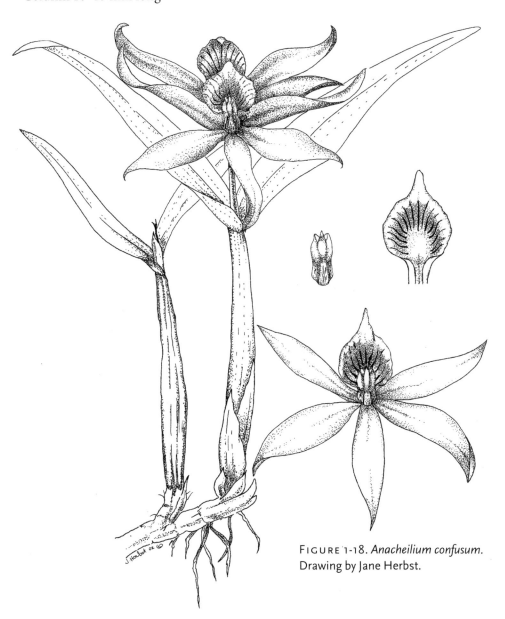

FIGURE 1-18. *Anacheilium confusum*. Drawing by Jane Herbst.

Anacheilium crassilabium
FIGURE 1-19, PLATES 20, 21, 22

Anacheilium crassilabium (Poeppig & Endlicher) Withner, Harding & Campacci, *comb. nov.* Basionym: *Epidendrum crassilabium* Poeppig & Endlicher. 1838. *Nov. Gen. Sp.* 2: 1. t. 102.

SYNONYMS

Epidendrum variegatum W. J. Hooker. 1832. *Bot. Mag.* 59: t. 3151.

Epidendrum coriaceum Parker *ex* W. J. Hooker. 1837. *Bot. Mag.* 64: t. 3595.

Epidendrum crassilabium Poeppig & Endlicher. 1838. *Nov. Gen. Sp.* 2: 1. t.102.

Epidendrum variegatum var. *virens* Lindley. 1853. *Folia Orchidacea Epidendrum* 38.

Epidendrum variegatum var. *coriaceum* (Parker *ex* Hooker) Lindley. 1853. *Folia Orchidacea Epidendrum* 38.

Epidendrum variegatum var. *crassilabium* (Poeppig & Endlicher) Lindley. 1853. *Folia Orchidacea Epidendrum* 38.

Epidendrum variegatum var. *leopardinum* Lindley. 1853. *Folia Orchidacea Epidendrum* 38.

Epidendrum coriaceum Focke. 1853. *Botanische Zeitung (Berlin)* 11: 228.

Epidendrum pachysepalum Klotzsch. 1855. *Allgemeine Gartenzeitung* 23: 274.

Epidendrum variegatum var. *lineatum* Reichenbach f. 1856. *Bonplandia* 4: 326.

Epidendrum christii Reichenbach f. 1877. *Linnaea* 41: 112.

Epidendrum leopardinum Reichenbach f. 1877. *Linnaea* 41: 112.

Epidendrum longipes Reichenbach f. 1878. *Otia Bot. Hamburgensia* 10.

Aulizeum variegatum (W. J. Hooker) Lindley *ex* Stein. 1892. *Orchideenbuch* 241.

Epidendrum feddeanum Kraenzlin. 1905. *Repertorium Specierum Novarum Regni Vegetabilis* 1: 188.

Epidendrum saccharatum Kraenzlin. 1908. *Orchis* 2: 113, f.17.

Epidendrum rhabdobulbon Schlechter. 1920. *Repertorium Specierum Novarum Regni Vegetabilis, Beihefte.* 7: 146.

Epidendrum baculibulbum Schlechter. 1923. *Repertorium Specierum Novarum Regni Vegetabilis, Beihefte* 19: 116.

Epidendrum rhopalobulbon Schlechter. 1924. *Repertorium Specierum Novarum Regni Vegetabilis, Beihefte* 27: 72.

Auliza wilsoni Galé. 1938. *Cat. Descriptivo de las Orquideas Cubanas* 60: 85.

Epidendrum variegatum var. *angustipetalum* Hoehne. 1947. *Arquivos de botanica do estado de São Paulo*, n.s., 2: 82.

Encyclia crassilabia (Poeppig & Endlicher) Lemée. 1955. *Flore de la Guyane Française* 1: 418.

Hormidium variegatum (W. J. Hooker) Brieger. 1960. *Publicacao Cientifica Universidade se São Paulo, Institut de Genetica*, 1: 20.

Hormidium coriaceum (Parker) Brieger. 1960. *Publicacao Cientifica Universidade se São Paulo, Institut de Genetica*, 1: 20.

Encyclia crassilabia (Poeppig & Endlicher) Dressler. 1961. *Brittonia* 13: 264.

Hormidium virens (Lindley) Brieger. 1961. *Publicacao Cientifica Universidade de São Paulo, Institut de Genetica*, 2: 69.

Encyclia vespa (Vellozo) Dressler. 1971. *Phytologia* 21: 441; isonym: *Encyclia vespa* (Vellozo) Pabst. 1967–1972. *Orquídea* 29 (6): 277.

Hormidium lineatum (Reichenbach f.) Brieger. 1977. Schlechter's *Die Orchideen*, ed. 3, 571.

Encyclia leopardina (Reichenbach f.) Dodson & Hágsater. 1994. *Orquideología* 19 (2): 149.

Prosthechea vespa (Vellozo) W. E. Higgins. 1997. *Phytologia* 82 (5): 381.

Prosthechea leopardina (Reichenbach f.) Dodson & Hágsater. 1999. *Monographs in Systematic Botany* 75: 956.

Prosthechea christii (Reichenbach f.) Dodson & Hágsater. 1999. *Monographs in Systematic Botany* 75: 956.

Prosthechea crassilabia (Poeppig & Endlicher) Carnevali & Ramirez. 2003. *Flora of the Venezuelan Guayana* 7: 538.

DERIVATION OF NAME

Latin *crasse*, "thick," and *labium*, "lip," referring to the thick lip.

DESCRIPTION

An epiphyte, terrestrial, or lithophyte. Rhizome segments not conspicuous. Pseudobulbs clustered, large, bimorphic, ovoid, compressed, or long and cylindric. Leaves two to four. Flowers five to fifteen, non-resupinate, fleshy, rigid, coloring varies, greenish white to brown with dark purple spots to transversely barred markings. Lip white with pink-purple markings on thickened apex, though this is also variable. Lip unlobed, or obscurely three-lobed, shortly clawed, adnate just above middle of the column, abruptly contracted to a fleshy acute, keeled apex, disc covered by a raised, fleshy, sulcate, soft, and folded callus conforming to the under surface of the column. Column pale green, this is also variable. Ovary three-winged.

HABITAT AND DISTRIBUTION

Costa Rica, Nicaragua, Panama, South America, and the West Indies. Found in pine and oak forest.

FLOWERING TIME
May and September to October.

CULTURE
Culture for this group is high light, lots of water when in growth, and intermediate conditions. We grow these in baskets hanging high in the greenhouse so they receive lots of light and air movement.

FIGURE 1-19. *Anacheilium crassilabium*. Drawing by Jane Herbst.

COMMENT

History has lumped many plants into this group, which used to be known as *Encyclia vespa*; however, the real *vespa* has been determined to be distinct from this group (see *Anacheilium vespa*). Similarly, *Anacheilium pamplonense* and *A. tigrinum* are also well differentiated. An article by Dunsterville and Dunsterville (1980) unambiguously separates them: those with a rounded lip are *A. tigrinum*, those with a quadrate lip are *A. pamplonense*, and those with a spadelike pointed lip are *A. crassilabium* and *A. vespa*. This separates out some of the variability; however, *A. crassilabium* still remains a highly variable species, in pseudobulb shape, flower color, and floral markings.

Plate 20, a picture of the type drawing, is a first step in sorting this puzzle. Plates 21 and 22 show two color forms of former "vespas" now *Anacheilium crassilabium*. Plate 50 shows a yet-to-be sorted out third form compared with *A. vespa*. One can see that these flowers are very different, but just how they differ and where to place the lines of division are not yet clear.

MEASUREMENTS

Pseudobulbs 6–20 cm long, 0.5–5 cm wide
Leaves 14–40 cm long, 2.5–5 cm wide
Inflorescence 30 cm long
Spathe 5 cm long
Sepals 10–13 mm long, 3–6 mm wide
Petals 7–13 mm long, 2.5–6 mm wide
Lip 7–8 mm long, 4–8 mm wide
Column 5 mm long

Anacheilium faresianum
PLATE 23

Anacheilium faresianum (Bicalho) Pabst, Bicalho, Moutinho & A. V. Pinto. 1981. *Bradea* 3: 23. Basionym: *Hormidium faresianum* Bicalho. 1973. *Bol. Soc. Camp. Orq.* 3 (3): 91. Type: Brazil, Minas Gerais, near Santo Antonio do Itambé, flowered in cultivation 10 July 1973, *Aniso Fares s.n.* (holotype: HB).

SYNONYMS

Encyclia faresiana (Bicalho) Pabst ex The International Plant Name Index.
Prosthechea faresiana (Bicalho) W. E. Higgins. 1997. *Phytologia* 82 (5): 381.

DERIVATION OF NAME

Honors Aniso Fares, the collector of the type specimen.

DESCRIPTION

A robust epiphyte. Pseudobulbs clustered, ovoid, slightly compressed. Leaves three. Flowers six, non-resupinate. Sepals and petals yellow-white with red venation, sepals with red color overlay (suffusion). Lip adnate to column to midway, rounded, sharply pointed at the apex, lip yellow with red oblong sulcate callus (a valley down the middle), the callus red becoming white toward the apex. Column short, thick with purple stripes, anthers yellow. Ovary three-winged or ridged, greenish purple.

HABITAT AND DISTRIBUTION

Brazil (Minas Gerais).

MEASUREMENTS

Pseudobulbs 7–9 cm long, 3 cm wide
Leaves 18–20 cm long, 3–3.5 cm wide
Inflorescence 15–17 cm long
Spathe 6 cm long
Sepals 19–20 mm long, 7 mm wide
Petals 19–20 mm long, 6 mm wide
Lip 15 mm wide

Anacheilium farfanii
FIGURE 1-20

Anacheilium farfanii (Christenson) Withner & Harding, *comb. nov.* Basionym: *Prosthechea farfanii* Christenson. 2002. *Orchids* 71 (8): 714. Type: PERU. Department of Cusco, Historic Sanctuary of Machu Picchu, Pampacahua, *Farfan s.n.* (holotype: CUZ).

DERIVATION OF NAME

Honors William Farfan Rios, an enthusiastic young student of orchids at the University of Cusco, Peru, and the discoverer of the plant.

DESCRIPTION

Pseudobulbs spindle-shaped, glossy. Leaves two. Inflorescence an erect raceme, with about five flowers. Flowers non-resupinate, shallowly cupped. Sepals and petals yellow heavily spotted and barred transversely with brown, the tips of the tepals yellow. Lip purple, unmarked, three-lobed, lateral lobes rhombic and rounded, strongly deflexed, midlobe roundly triangular, and pointed. Callus an inverted U-shaped ridge clasped by the column wings and extending beyond the column. Column with three pointed, toothlike projections around the anther, the uppermost with an additional thick superior blunt projection of equal length. Ovary three-angled.

HABITAT AND DISTRIBUTION
Endemic to southern Peru. Found at 2100–2600 meters in elevation. Grows on rocks in tufts.

FLOWERING TIME
February to March.

COMMENT
The species is similar to but differs from *Anacheilium hajekii* by having the markings on the sepals and petals distributed differently, forming broad concentric rings, and by having rounded lateral lobes on the lip and a shorter callus.

MEASUREMENTS
Pseudobulbs 12 cm long
Leaves 23 cm long, 2.3 cm wide
Inflorescence 9.5–13.2 cm long
Spathe 6 cm long
Sepals 11 mm long, 4–5 mm long
Petals 9 mm long, 4 mm wide
Lip 6 mm long, 6 mm wide
Column 4 mm long

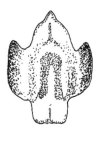

FIGURE 1-20. *Anacheilium farfanii*. Drawing by Jane Herbst.

Anacheilium faustum
PLATE 24

Anacheilium faustum (Reichenbach f. ex Cogniaux) Pabst, Moutinho & A. V. Pinto. 1981. *Bradea* 3: 23. Basionym: *Epidendrum faustum* Reichenbach f. ex Cogniaux. 1900. *Flora Brasiliensis (Martius)* 3 (5): 80.

SYNONYMS
Hormidium faustum (Reichenbach f.) Brieger. 1960. *Publicacao Cientifica Universidade se São Paulo, Institut de Genetica,*1: 20.
Encyclia fausta (Reichenbach f. ex Cogniaux) Pabst. 1967. *Orquídea* 29 (6): 276.
Prosthechea fausta (Reichenbach f. ex Cogniaux) W. E. Higgins. 1997. *Phytologia* 82 (5): 381.

DERIVATION OF NAME
After Faustus, a medieval magician who exchanged his soul for knowledge, power, and material gain.

DESCRIPTION
Pseudobulbs spindle-shaped, slightly compressed, on creeping rhizome separated by 1- to 1.5-cm segments. Leaves two or three. Inflorescence with four to ten starry white non-resupinate flowers. Flower 5 cm across, with pleasing astringent fragrance. Lip unlobed, markedly pointed with turned-down (revolute) sides and purple streaking. Column club-shaped.

HABITAT AND DISTRIBUTION
Brazil (Paraná, Santa Catarina, and Rio Grande do Sul). Found at 700 meters in elevation in areas of high humidity and daily misting or rain, cool to intermediate temperatures, and moderately bright light.

FLOWERING TIME
Summer to autumn in Brazil (Northern Hemisphere's spring and fall).

CULTURE
According to McQueen (1992, 77) this orchid grows well mounted or in baskets of coarsely chopped treefern or moss. It likes a cool to intermediate environment with moderate bright light.

COMMENT
We must admit that to identify this species through photographs is difficult, and it is easy to see not much difference between *Anacheilium faustum* and *A. gluma-*

ceum. Many photographs we received were labeled either *A. faustum* or *A. glumaceum* and we could not tell which was which. Plate 25 comes from Marcos Campacci in Brazil, and we believe his identification is correct. You must use the configuration of the lip to guide you.

MEASUREMENTS
Pseudobulbs 6–14 cm long, 0.6–1.3 cm wide
Leaves 12–20 cm long, 0.9–1.3 cm wide
Inflorescence 12 cm long
Spathe 6 cm long
Sepals 30–36 mm long, 6–10 mm wide
Petals 30–35 mm long, 9–10 mm wide
Lip 25–27 mm long, 11–13 mm wide
Column 7 mm long

Anacheilium fragrans
PLATE 25

Anacheilium fragrans (Swartz) Acuña Galé. 1938. *Cat. Orq. Cubanas* 86. Basionym: *Epidendrum fragrans* Swartz. 1788. *Prodr. Veg. Ind. Occ.* 123.

SYNONYMS
Epidendrum lineatum Salisbury. 1796. *Prodr. Stirp. Chap. Allerton* 10.
Epidendrum ionoleucum Hoffmannsegg ex Reichenbach f. 1852. *Linnaea* 25: 244.
Epidendrum fragrans var. *ionoleucum* Hoffmannsegg ex Barbosa Rodrigues. 1881. *Gen. Sp. Orch. Nov.* 2: 136.
Epidendrum fragrans var. *magnum* Stein. 1892. *Orchideenbuch* 230.
Epidendrum vaginatum Sessé & Mociño. 1894. *Fl. Mex.*, ed. 2: 201.
Epidendrum fragrans var. *pachypus* Schlechter. 1922. *Repertorium Specierum Novarum Regni Vegetabilis* 17: 32.
Encyclia fragrans (Swartz) Lemée. 1955. *Flore de la Guyane Française* 1: 418.
Hormidium fragrans (Swartz) Brieger. 1961. *Publicacao Cientifica Universidade de São Paulo, Institut de Genetica*, 2: 69.
Prosthechea fragrans (Swartz) W. E. Higgins. 1997. *Phytologia* 82 (5): 381.

DERIVATION OF NAME
Latin *fragrans*, "scented."

DESCRIPTION

An epiphyte. Pseudobulbs variable in shape, on tough rhizome at intervals 1–4 cm apart. Leaf one or occasionally two. Flowers three to ten, non-resupinate. Sepals and petals creamy white, sometimes with fine purple spots on back of sepals. Lip unlobed, strongly concave with point at apex, white to cream with faint to strong purple veining. Column cream with green at base, brownish cream underneath, sparsely marked with purple. Column three-toothed. Anther yellow. Capsule three-winged.

HABITAT AND DISTRIBUTION

Central America and West Indies with one record from Guyana. Found at 50–2000 meters in elevation.

FLOWERING TIME

May to June and sporadically into August.

CULTURE

Culture for these plants is intermediate to warm temperatures, watering throughout the year, with daily watering in the spring and summer, as long as the mix does not stay wet. The plants grow in areas of high humidity and almost daily rain. They will tolerate long periods (weeks) of drying out, but would prefer to be watered daily, having the roots dry slightly between watering. They can be mounted or potted, making sure that if mounted they are watered enough so the roots do not dry too much. These too have a wonderful fragrance and blooms that last a long time.

COMMENT

Confusion exists concerning *Epidendrum lineatum* Salisbury (1796, *Prodr. Stirp. Chap. Allerton* 10), based on *Curtis Botanical Magazine* plate 152 (as "*Epidendrum cochleatum*") from Jamaica. As this plant is from Jamaica, it is a synonym of *Anacheilium fragrans* and not *A. aemulum* as some sources report.

MEASUREMENTS

Pseudobulbs 4.5–11 cm long, 1–3 cm wide
Leaves 9–31 cm long, 1.2–4.8 cm wide
Inflorescence 5–17 cm long
Spathe 4 cm long
Sepals 20–35 mm long, 4–9 mm wide
Petals 20–30 mm long, 7–12 mm wide
Lip 17–24 mm long, 9–17 mm wide
Column 6.5–8 mm long

Anacheilium fuscum

FIGURE 1-21, PLATE 26

Anacheilium fuscum (Schlechter) Withner & Harding, *comb. nov.* Basionym: *Epidendrum fuscum* Schlechter. 1921. *Repertorium Specierum Novarum Regni Vegetabilis, Beihefte* 9: 84. Holotype: Peru, Cajamarca, eastern shelf of the Cordillera above Tabaconas, 2400 meters, May 1912, *Weberbauer 6301* (holotype: B, destroyed); neotype: PERU: Cusco, road to Quillabamba, ca. 2500 m, May 1995, *M. Cavero & J. Leon G. 1664* (neotype: USM; isoneotype: MOL).

SYNONYMS

Encyclia fusca (Schlechter) D. E. Bennett & Christenson. 1998. *Icones Orchidacearum Peruvianum*, pl. 442.

Prosthechea fusca (Schlechter) D. E. Bennett & Christenson. 2001. *Icones Orchidacearum Peruvianum*, pl. 750.

DERIVATION OF NAME

Latin *fuscus*, "somber brown" or as Lindley put it "brown tinged with grayish or blackish," referring to the color of the flowers.

DESCRIPTION

An epiphyte with a creeping rhizome. Pseudobulbs spindle-shaped, slender. Leaves two or three. Flowers eight to ten, non-resupinate. Sepals and petals externally pale cream-yellow with pale lavender suffusion at base and apex, internally dark brownish-rose with a narrow pale green band across the base. Lip dark purplish-rose, the nipplelike apex cream-yellow. Callus pink or rose with warty trichomes.

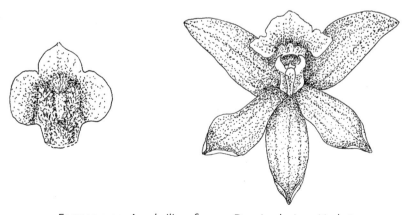

FIGURE 1-21. *Anacheilium fuscum*. Drawing by Jane Herbst.

Column pale green with faint minute brownish-rose spots. Lip three-lobed, adnate only at base, lateral lobes rhombic-suborbicular, obtuse rounded, midlobe transverse, hemispheric, obtuse. Callus a densely pubescent (hairy) rectangular pad.

HABITAT AND DISTRIBUTION
Peru. Found at 2400 meters in elevation.

FLOWERING TIME
April to May.

COMMENT
Anacheilium fuscum and *A. bennettii* are very similar and it is difficult to tell them apart without holding the actual plant in your hand. Many plants over the years have been misidentified as *A. fuscum* when they properly should have been *A. bennettii*.

Plate 26, of *Anacheilium fuscum*, was taken by Carlos Hajek of Peru, and though it looks like it could be *A. hartwegii*, because of the country of origin it is more likely *A. fuscum*. Further research may show them to be the same species.

MEASUREMENTS
Pseudobulbs 15 cm long
Leaves 23–25 cm long, 2.9–3.1 cm wide
Inflorescence 23 cm long
Sepals 14 mm long, 6 mm wide
Petals 12 mm long, 6 mm wide
Lip 7–8 mm long, 10 mm wide
Column 6 mm long

Anacheilium garcianum
FIGURE 1-22, PLATE 27

Anacheilium garcianum (Garay & Dunsterville) Withner & Harding, *comb. nov.* Basionym: *Epidendrum garcianum* Garay & Dunsterville. 1961. *Venezuelan Orchids Illustrated* 2: 122.

SYNONYMS
Encyclia garciana (Garay & Dunsterville) Carnevali & I. Ramírez. 1986. *Ernestia* 36: 9.
Prosthechea garciana (Garay & Dunsterville) W. E. Higgins. 1997. *Phytologia* 82 (5): 381.

DERIVATION OF NAME
Honors Carlos García Esquivel, M.D., eminent Venezuelan orchidist.

DESCRIPTION
An epiphyte. Pseudobulbs variable, compressed, light green with finely wrinkled surface. Leaf one. Inflorescence of single flower, non-resupinate from long loose fitting sheath. Sepals and petals rather fleshy, white overlaid in varying intensity

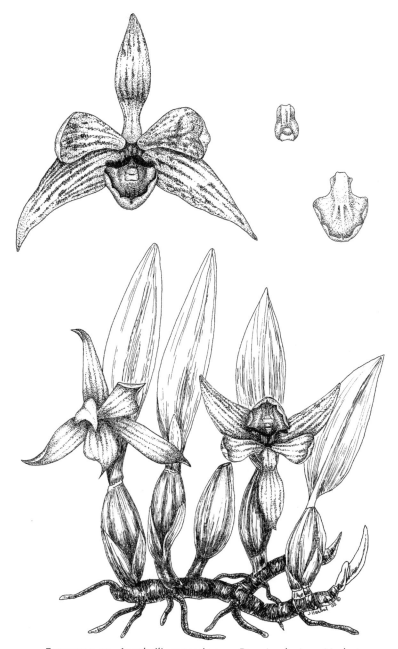

FIGURE 1-22. *Anacheilium garcianum.* Drawing by Jane Herbst.

with plum-colored nerves and suffusion, the backs tend to be paler than the faces. Lip three-lobed, smaller lateral lobes with midlobe larger and pointed, white sometimes with pale plum nerve lines. Upraised callus at base is hollow underneath, upper surface glandular grading into short white hairs thickly covering apical portion. Column three-toothed, anther cream. Ovary sharply triangular in cross section.

HABITAT AND DISTRIBUTION
Venezuela. Found at 1220 meters in cloud forest.

FLOWERING TIME
October.

CULTURE
Culture for this plant is intermediate with high humidity and frequent watering. We grow this in a basket, which gets water daily in the spring and summer and weekly in the fall and winter.

MEASUREMENTS
Pseudobulbs 4 cm long, 1.2 cm wide
Leaves 5–9 cm long, 1.5–2 cm wide
Inflorescence (peduncle and ovary) 4 cm long
Spathe 2 cm long
Sepals 27 mm long, 10 mm wide
Petals 22 mm long, 12 mm wide
Lip 18 mm long, 12 mm wide
Column 7 mm long

Anacheilium gilbertoi
PLATE 28

Anacheilium gilbertoi (Garay) Withner & Harding, *comb. nov.* Basionym: *Epidendrum gilbertoi* Garay. 1971. *Orquideología* 6: 16. Type: Colombia, Department of Caldas, Anserma, 2000 meters, *G. Escobar 586* (holotype: AMES).

SYNONYMS
Encyclia gilbertoi (Garay) P. Ortiz. 1991. *Orquideología* 18 (1): 99.
Prosthechea gilbertoi (Garay) W. E. Higgins. 1997. *Phytologia* 82 (5): 381.

DERIVATION OF NAME
Honors Gilberto Escobar, who collected this plant.

DESCRIPTION
An epiphyte 40 cm tall. Rhizome ascending, moderately thick and covered with a withered sheath. Pseudobulbs cylindrical to spindle-shaped. Leaves two, leathery. Flowers few, fleshy. Tepals orange to deep red; lip white with red veins, chocolate-brown coloration medially. Lip concave, heart-shaped at base, margins crisped. Disc thickened, strongly convex, cushionlike, three-lobed at base of callus extending to midway of lip. Column two-horned.

HABITAT AND DISTRIBUTION
Colombia. Found at 2000 meters in elevation.

CULTURE
J. Bayron Pineda Palacio of Medellín, Colombia, gave Patricia a piece of his plant, which has deep red flowers. He says to grow it cool, fertilize it once a month, and mount it on tree fern. Jim Figura in New York grows his variety with orange flowers in warm conditions and he says it rambles about on flat surfaces. Ask your supplier the elevation or climate type of the plant's origin if you want to grow this plant properly.

COMMENT
Anacheilium gilbertoi is similar to *A. lambda* but the later is smaller and has a triangular cordate lip with distinct apiculate (pointed) tip.

MEASUREMENTS
Pseudobulbs 15 cm long, 0.5 cm wide
Leaves 20 cm long, 2 cm wide
Inflorescence 10 cm long
Sepals 14 mm long, 3 mm wide
Petals 13 mm long, 3 mm wide
Lip 11 mm long, 13 mm wide

Anacheilium glumaceum
FIGURE 1-23, PLATE 29

Anacheilium glumaceum (Lindley) Pabst, Moutinho & A. V. Pinto. 1981. *Bradea* 3 (23): 183. Basionym: *Epidendrum glumaceum* Lindley. 1839. *Edward's Botanical Register* 25, misc. 38.

SYNONYMS

Aulizeum glumaceum (Lindley) ex Stein. 1892. *Orchideenbuch* 231.

Epidendrum almasii Hoehne. 1947. *Arquivos de botanica do estado de São Paulo*, n.s., 2: 84, t. 25.

Hormidium glumaceum (Lindley) Brieger. 1960. *Publicacao Cientifica Universidade de São Paulo, Institut de Genetica*, 1: 19.

Encyclia almasii (Hoehne) Pabst. 1967. *Orquídea* 29 (6): 276.

Encyclia glumacea (Lindley) Pabst. 1972. *Orquídea* 29 (6): 276.

Prosthechea glumacea (Lindley) W. E. Higgins. 1997. *Phytologia* 82: 378.

DERIVATION OF NAME

Latin *glume*, "chaffy in texture," referring to the chafflike bracts (spathe) enclosing the floral parts on grass.

DESCRIPTION

Pseudobulbs flat and narrow, widest below midsection. Leaves two. Spathe (glume) surrounding peduncle, brown and prominent. Flowers white tinged with pink, striped with rose. Lip convex not concave, obovate and pointed, widest below midsection, becoming pointed at apex.

HABITAT AND DISTRIBUTION

Brazil (Espírito Santo, Pernambusco, Paraná, Rio Grande do Sul, Santa Catarina, São Paulo), Ecuador, Guatemala, and Nicaragua. Found at 50 meters.

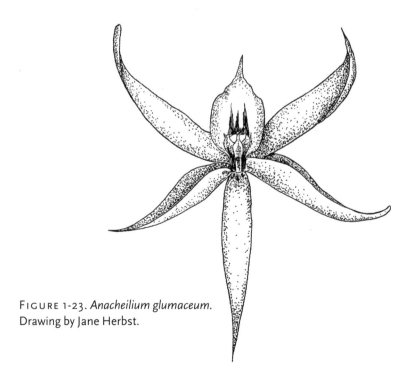

FIGURE 1-23. *Anacheilium glumaceum*. Drawing by Jane Herbst.

COMMENT

We repeat here that we found it difficult to differentiate this species from *Anacheilium faustum* by photograph alone.

Epidendrum almasii Hoehne is included here as a probable synonym, although Pabst at times handles it as a separate species. The spelling is variable in the literature, and we have seen both *almasyi* and *almasii*. We present this name for those wishing to differentiate these plants from the broadly defined *Anacheilium glumaceum*. They are found in Brazil, Sierra del Mar to Rio de Janeiro and Paraná and Mantiqueira, also the south of Minas Gerais, and common near São Paulo.

The pseudobulbs are spindle-shaped, flattened, two-sided, base extremely narrowed. Leaves two or three. The raceme has eight to fifteen large flowers, which smell like roses. Sepals and petals are pointed, creamy color with short reddish lines at base of the lip and sometimes at the inside base of the petals. Plants resembling *Epidendrum almasii* are described as being like *Anacheilium glumaceum* but bigger and more ornamental, but Hoehne states that the presentation of the flowers is different.

Epidendrum almasii measurements: pseudobulbs 8–18 cm long, 2–4.5 cm wide; leaves 12–20 cm long, 2–4 cm wide; sepals 30–36 mm long, 6–8 mm wide; petals 25–28 mm long, 6–10 mm wide; and lip 15–18 mm long, 9–15 mm wide.

MEASUREMENTS

Pseudobulbs 5–10 cm long, 1–2 cm wide
Leaves 10–20 cm long, 2–3 cm wide
Inflorescence 15 cm long
Sepals 20–26 mm long, 3–4 mm wide
Petals 15–17 mm long, 4 mm wide
Lip 14–15 mm long, 7–8 mm wide
Column 4–5 mm long

Anacheilium hajekii
FIGURE 1-24, PLATE 30

Anacheilium hajekii (Bennett & Christenson) Withner & Harding, *comb. nov.* Basionym: *Prosthechea hajekii* D. E. Bennett & E. A. Christenson. 2001. *Icones Orchidacearum Peruvianum*, pl. 750. Type: Peru: Junin: Chanchamayo, up the steep canyon oriented NW of the road, 8 kilometers before San Ramon, 1800 meters, 11 November 1998, *C. Hajek & F. Hajek ex Bennett 7938* (holotype: MOL).

The Genus *Anacheilium*

DERIVATION OF NAME
Honors Carlos and Frank Hajek.

DESCRIPTION
Mat-forming terrestrial up to 40 cm tall. Pseudobulbs slender, spindle-shaped, slightly compressed. Leaves two. Flowers to 20. Sepals and petals greenish yellow

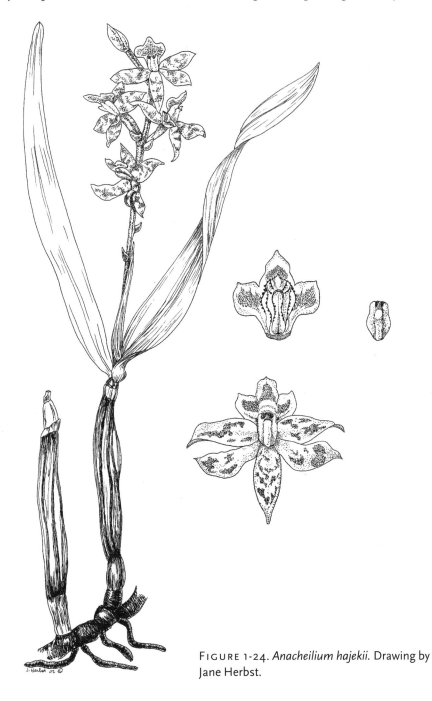

FIGURE 1-24. *Anacheilium hajekii*. Drawing by Jane Herbst.

with the inner surface having brown to purple blotches. Lip three-lobed, lateral lobes suborbicular with nipplelike points, midlobe rounded with broadly triangular apex, cream white with lavender suffusion over most of the surface. Callus thick, form fitting with the column, oblong with central oblong lobe. Column green at base becoming pale cream-yellow at apex. Anther orange-yellow.

HABITAT AND DISTRIBUTION
Peru. Found at 1600–1800 meters, in lower levels of cool montane cloud forest. Carlos Hajek (pers. comm.) sends this habitat information: wet montane forest, lots of rain during December to April (summer), elevation 1600 meters, temperature 14–28°C.

FLOWERING TIME
Late July to August.

CULTURE
Culture on this one is unknown.

COMMENT
This species is like *Anacheilium hartwegii* but is distinguished by the pointed dorsal sepal, nipplelike points on the lateral lobes of lip, and blotched sepals and petals.

MEASUREMENTS
Pseudobulbs 11–15 cm long, 0.8–1 cm wide
Leaves 15–23 cm long, 1.5–2 cm wide
Inflorescence 8.5 cm wide
Sepals 11 mm long, 5 mm wide
Petals 9 mm long, 3.5 mm wide
Lip 6.5 mm long, 6.5 mm wide
Column 5.5 mm long, 3.5 mm wide

Anacheilium hartwegii
FIGURE 1-25, PLATE 31

Anacheilium hartwegii (Lindley) Pabst, Moutinho & A. V. Pinto. 1981. *Bradea* 3: 23. Basionym: *Epidendrum hartwegii* Lindley. 1844. *Bentham Pl. Hartweg.* 150.

SYNONYMS
Encyclia hartwegii (Lindley) R. Vásquez & Dodson. 1982. *Icones Plantarum Tropicarum* 527.
Prosthechea hartwegii (Lindley) W. E. Higgins. 1997. *Phytologia* 82 (5): 381.

DERIVATION OF NAME

Honors German botanist Karl Theodor Hartweg (1812–1871).

DESCRIPTION

An epiphyte with a creeping rhizome. Pseudobulbs spindle-shaped, compressed, old pseudobulbs lightly grooved. Leaves two. Flowers five to eight, non-resupinate. Sepals and petals green with maroon, blood-color markings. Sepals very thick with square edged margins. Petals not as thick as sepal margins, not so square-edged. Shape very irregular. Lip three-lobed, adnate only at base, thick, rigid; lateral lobes

FIGURE 1-25. *Anacheilium hartwegii*. Drawing by Jane Herbst.

subequaling midlobe. Lip white with one to several purple marks and pale green at base. Callus covered with very short glandular hairs. Column white with strongly pronounced dorsal ridge, anther yellow, three-toothed. Ovary three-winged.

HABITAT AND DISTRIBUTION
Ecuador. Found at 1300–3000 meters.

CULTURE
Culture is intermediate to warm, wet in the spring and summer, slightly drier in the fall and winter.

COMMENT
Eric Christenson (pers. comm.) doubts that *Anacheilium hartwegii* has sepals and petals that are marked. He suspects they are brown like those of *A. fuscum*. The misunderstanding occurs because *A. hartwegii* and *A. brachychilum* have been confused, and the written type descriptions do not mention color. Use the shape of the lip to identify these plants.

MEASUREMENTS
Pseudobulbs 10–17 cm long, 2 cm wide
Leaves 16 cm long, 2 cm wide
Inflorescence 15 cm long
Sepals 7–8 mm long, 3.7–4 mm wide
Petals 8 mm long, 5 mm wide
Lip 5.5 mm long, 5.5 mm wide
Column 3.5 mm long, 4 mm wide

Anacheilium ionophlebium
PLATE 32

Anacheilium ionophlebium (Reichenbach f.) Withner & Harding, *comb. nov.* Basionym: *Epidendrum ionophlebium* Reichenbach f. 1866. *Beitr. Orch. Centr. Am.* 103.

SYNONYMS
Epidendrum madrense Schlechter. 1918. *Beih. Bot. Centralbl.* 36 (2): 405.
Epidendrum hoffmanii Schlechter. 1920. *Repertorium Specierum Novarum Regni Vegetabilis* 16: 444.
Encyclia ionophlebia (Reichenbach f.) Dressler. 1961. *Brittonia* 13: 264.
Prosthechea ionophlebia (Reichenbach f.) W. E. Higgins. 1997. *Phytologia* 82 (5): 381.

DERIVATION OF NAME

Greek *iono*, "violet-colored," and *phlebius*, "veined," referring to the lip markings.

DESCRIPTION

Pseudobulbs ovoid to spindle-shaped, compressed. Leaves two. Flowers two to seven, non-resupinate, about 6 cm in diameter, greenish yellow or greenish cream-colored. Lip unlobed, concave, with shallow notch at apex, with dull purplish radiating streaks. Callus simple, pubescent. Ovary three-winged.

HABITAT AND DISTRIBUTION

México to Panama, and Venezuela. Found at 700–1100 meters.

FLOWERING TIME

Winter to spring. We have also seen listed bloom times from March to June.

COMMENT

Anacheilium ionophlebium can be differentiated from *A. chacaoensis* by the notch at the lip apex and by the not fleshly thickened lip, which is usually marked with purple lines. Some authors consider these two species to be synonyms.

MEASUREMENTS

Pseudobulbs 3–14 cm long, 1.5–3 cm wide
Leaves 35 cm long, 4 cm wide
Inflorescence 15 cm long
Sepals 15–26 mm long, 3.5–8 mm wide
Petals 15–25 mm long, 5–8.5 mm wide
Lip 10–23 mm long, 10–20 mm wide
Column 7–11 mm long

Anacheilium janeirense

FIGURE 1-26, PLATE 33

Anacheilium janeirense Campacci, *sp. nov.*
   Specii Anacheilium vespa similis, sed herba majoris; rhizomate breve; pedunculus breve; petalae et sepalae brunneae, vitellinae marginatae; labellum breve et latum, luteo-album, purpurato maculato; etiam columna brevis et lata; purpurata maculata in basis. Type: Brazil, State of Minas Gerais, Barbacena city surroundings, 700–800 meters in elevation, February 1992, *Hermann Kundegraber, s.n.* (holotype: SP).

DERIVATION OF NAME

After the state of Rio de Janeiro, one of the places where the plant grows.

DESCRIPTION

An epiphyte. Rhizome segments short; roots glabrous, white. Pseudobulbs spindle-shaped, compressed laterally, apex tapered, covered by imbricating deciduous sheaths, 10–15 cm long and 2–3 cm wide. Leaves two or three, obtuse, 15–20 cm long and 2.5–3 cm wide. Inflorescence on erect pedunculate raceme emerging from the apex of the pseudobulbs, 15 cm long, floral bracts minute. Flowers eight to fifteen. Dorsal sepal lanceolate, acute apex, 1.8 cm long and 0.7 cm wide. Lateral sepals elliptic, oblique, apex acute, 1.8 cm long and 0.9 cm wide. Petals lanceolate, oblique, apex acute, 1.8 cm long and 0.6 cm wide. Lip elliptic, hard thick and glabrous at the base, 1 cm long and 0.8 cm wide, trilobulate, lateral lobes small, acute; callus conforming to the column, with two symmetrical lamellae congruent to the base. Sepals and petals dark brown with greenish-yellow border; lip yellowish with a few purple spots; callus whitish. Column gibbous, without auricles, 0.8 cm long and 0.4 cm wide. Anther subcircular, retuse, yellowish. Pollinarium with four yellowish pollinia. Stigmatic cavity cordiform.

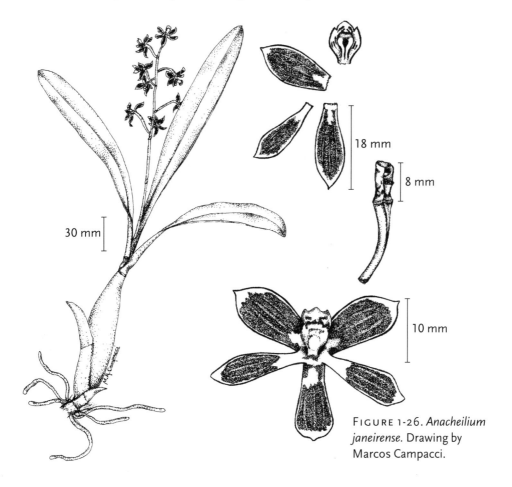

FIGURE 1-26. *Anacheilium janeirense*. Drawing by Marcos Campacci.

HABITAT AND DISTRIBUTION
Atlantic forest of Brazil. Found at 700–800 meters in elevation.

FLOWERING TIME
In fall in habitat.

COMMENT
This species has been found at many sites in the Atlantic forest of Brazil, occurring in the states of Minas Gerais, Espírito Santo, and Rio de Janeiro. The plant we refer to comes to us by way of Marcos Campacci, who did the drawing, wrote the description, and provided the picture.

MEASUREMENTS
Pseudobulbs 10–15 cm long, 2–3 cm wide
Leaves 15–20 cm long, 2.5–3 cm wide
Inflorescence 15 cm long
Sepals 18 mm long, 9 mm wide
Petals 18 mm long, 6 mm wide
Lip 10 mm long, 8 mm wide
Column 8 mm long, 4 mm wide

Anacheilium jauanum
FIGURE 1-27

Anacheilium jauanum (Carnevali & I. Ramírez) Withner & Harding, *comb. nov.* Basionym: *Encyclia jauana* Carnevali & Ramírez. 1994. *Lindleyana* 9 (1): 67. Type: Venezuela. Bolívar: Meseta de Jauna, Cerro Jaua, summit, portion SO, 4°48'50"N 64°34'10"W gallery forest of a tributary of the river Marajano, 1750–1800 meters, 22–28 February 1974, *Steyermark, Carreño-Espinoza & Brewer-Carías 109478* (holotype: VEN; fragment, AMES).

SYNONYM
Prosthechea jauana (Carnevali & I. Ramírez) W. E. Higgins. 1997. *Phytologia* 82
 (5): 381.

DERIVATION OF NAME
After Cerro Jaua in Venezuela, where the plant grows.

DESCRIPTION
A sun-loving epiphyte to 11.6 cm tall. Rhizome thick, short to fairly well developed with pseudobulbs 2.3 cm apart. Pseudobulbs ellipsoid, clothed by sheaths that

envelope the basal half and eventually fall off. Inflorescence two- to five-flowered. Flowers non-resupinate, with widely spreading creamy white perianth segments. Lip white with 15–17 longitudinal purple nerves (innermost nerves bifurcate at apex). Lip fused with column 2–3 mm from base, concave in natural position, ovate when flattened, abruptly to point at apex. Callus oblong-subquadrate, slightly depressed at center. Column with three teeth. Flowers from mature growths.

HABITAT AND DISTRIBUTION
Venezuela from the Meseta de Jauna, a flat-topped tepui. Found at 1400–1800 meters, in dry climates sometimes with prolonged dry periods.

COMMENT
This species is similar to *Anacheilium fragrans* except the stigma is ovate-cordate (rounded heart-shaped), the callus is slightly depressed, the column has a deeper clinandrium with shorter lateral teeth, and the two species are geographically isolated (*Anacheilium fragrans* does not occur in Venezuela).

MEASUREMENTS
Pseudobulbs 2.5–4.5 cm long, 1.2–1.5 cm wide
Leaves 3.3–7.5 cm long, 1.3–2.2 cm wide
Inflorescence 7 cm long
Spathe 2.3–2.5 cm long
Sepals 19–21 mm long, 5–6 mm wide
Petals 17–19 mm long, 5.5–6 mm wide
Lip 12–14 mm long, 11–13 mm wide
Column 8 mm long, 4 mm wide

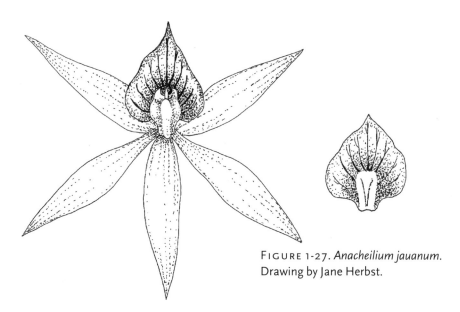

FIGURE 1-27. *Anacheilium jauanum*. Drawing by Jane Herbst.

Anacheilium joaquingarcianum

Anacheilium joaquingarcianum (Pupulin) Withner & Harding, *comb. nov.*
Basionym: *Prosthechea joaquingarciana* Pupulin. 2001. *Selbyana* 22 (1): 19.
Type: Costa Rica. Heredia: Varablanca, ca. 1800 meters, collected by Zayda Rodríguez, 1997, flowered in cultivation at El Roble de Alajuela, 4 December 1999, *F. Pupulin 1865* (holotype: USJ).

DERIVATION OF NAME
Honors Joaquín B. García Castro, a student of Costa Rican orchid flora.

DESCRIPTION
An epiphyte. Pseudobulbs cylindric, linear to subfusiform (long spindle-shaped). Leaves two. Inflorescence a raceme of eight flowers. Sepals and petals brownish yellow, spotted with brown, lip white and striped with purple. Lip ovate, concave toward the apex. Disk of lip composed of two calli formed by two membranous keels, which unite centrally and diverge at apex of disk. Column dilated at the middle, with two lateral fleshy teeth at apex and a central 2–dentate projection.

HABITAT AND DISTRIBUTION
Costa Rica. Found at 1800–2000 meters in elevation, in lower montane and montane wet forests.

MEASUREMENTS
Pseudobulbs 24 cm long, 1.2 cm wide
Leaves 18–22 cm long, 2.2–2.5 cm wide
Inflorescence 19 cm long
Sepals 20 mm long, 7 mm wide
Petals 16 mm long, 7.5 mm wide
Lip 13.5 mm long, 11.2 mm wide
Column 10 mm long

Anacheilium kautskyi
PLATE 34

Anacheilium kautskyi (Pabst) Pabst, Moutinho & A. V. Pinto. 1981. *Bradea* 3: 23. Basionym: *Encyclia kautskyi* Pabst. 1967. *Orquídea* 29: 63, tab. 3. Type: Brazil, near Domingos Martins (holotype: HB 20.552).

SYNONYM

Prosthechea kautskyi (Pabst) W. E. Higgins. 1997. *Phytologia* 82 (5): 381.

DERIVATION OF NAME

Honors Roberto Kautsky.

DESCRIPTION

An epiphyte, forming dense mats with thick horizontal leads. Roots thick, white, forming an extensive system that is not very adherent to the substrate. Pseudobulbs spaced 2 cm apart on a thick tough rhizome, pseudobulbs ovoid, compressed. Leaves two, stiff. Flowers to six, non-resupinate, yellowish brown. Sepals and petals sharply pointed. Lip projecting, heart-shaped, whitish with nine purple stripes.

HABITAT AND DISTRIBUTION

Brazil (Espírito Santo). Found at 580–1200 meters.

CULTURE

Culture for this plant is warm, wet, and humid.

COMMENT

Several pictures for this plant were submitted to us, some not even close to correct, and others that were *Anacheilium campos-portoi*. The number of leaves seems to be a quick way to separate *A. kautskyi* (two leaves) from *A. campos-portoi* (one leaf).

We are thankful to Vitorino Paiva Castro Neto for his help in obtaining the photograph from Mr. Kautsky, and to Roberto Kautsky for supplying the photograph.

MEASUREMENTS

Pseudobulbs 2.5–4 cm long, 0.5 cm wide
Leaves 3–4.5 cm long, 1–1.8 cm wide
Inflorescence 2–4 cm long
Spathe 1.5 cm long
Sepals 8 mm long, 2.5–2.8 mm wide
Petals 7 mm long, 0.8 mm wide

Anacheilium lambda
FIGURE 1-28

Anacheilium lambda (Linden & Reichenbach f.) Withner & Harding, *comb. nov.* Basionym: *Epidendrum lambda* Linden & Reichenbach f. 1854. *Bonplandia* 2: 281.

SYNONYMS

Epidendrum rueckerae Reichenbach f. 1865. *Hamburger Garten- und Blumenzeitung* 21: 385.

Encyclia lambda (Linden & Reichenbach f.) Dressler. 1971. *Phytologia* 21: 440.

Prosthechea lambda (Linden & Reichenbach f.) W. E. Higgins. 1997. *Phytologia* 82 (5): 381.

DERIVATION OF NAME

Flower parts have yellow lines forked and flecked to make little lambdas, hence the species name.

DESCRIPTION

Pseudobulbs compressed, spindle-shaped, with sheath enclosing bulb. Leaves two. Inflorescence a many-flowered raceme. Tepals subequal, basally narrow, broadening to a wedgelike taper. Lip heart-shaped at the base, triangular at the apex, pointed, rayed obscurely with violet-purple veins that radiate at apex, some becoming slightly forked. Callus depressed, well rounded, with lateral longitudinal purple lines on both sides, pubescent. Column thick, basally attached to lip, apex three-toothed, lateral teeth acute, medial tooth rounded. Flowers are very fragrant.

HABITAT AND DISTRIBUTION

Colombia (Ocaña). Found at 200–250 meters in elevation.

FLOWERING TIME

June.

COMMENT

Dressler (1971) suggests this may be a natural hybrid between *Anacheilium baculus* and *A. chacaoense*.

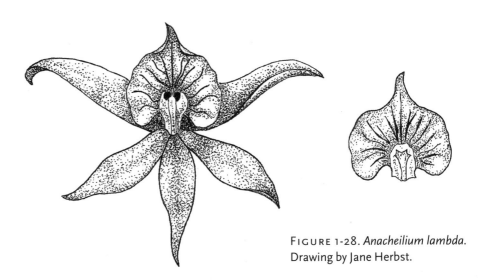

FIGURE 1-28. *Anacheilium lambda*. Drawing by Jane Herbst.

Anacheilium lindenii

FIGURE 1-29, PLATE 35

Anacheilium lindenii (Lindley) Withner & Harding, *comb. nov.* Basionym: *Epidendrum lindenii* Lindley. 1843. *Ann. Mag. Nat. Hist.* 12: 397.

SYNONYMS

Epidendrum fallax Lindley. 1846. *Orchidaceae Lindenianae* 9.
Epidendrum fallax b. flavescens Lindley. 1853. *Folia Orchidacea Epidendrum* 35.
Epidendrum fallax var. *flavescens* Reichenbach f. 1854. *Bonplandia* 2: 20.
Epidendrum flavescens Schlechter. 1919. *Repertorium Specierum Novarum Regni Vegetabilis, Beihefte* 6: 69.
Prosthechea lindenii (Lindley) W. E. Higgins. 1997. *Phytologia* 82 (5): 381.

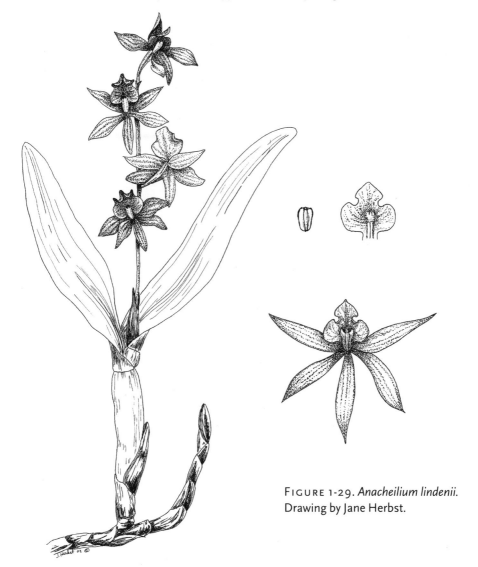

FIGURE 1-29. *Anacheilium lindenii*. Drawing by Jane Herbst.

DERIVATION OF NAME

Honors Jean Linden (1817–1898), a Belgian horticulturist, who collected the type specimen.

DESCRIPTION

An epiphyte with a thick tough rhizome clothed in brown sheaths that last for about two years. Pseudobulbs spaced 3–8 cm apart, variable in shape, moderately compressed. Inflorescence of five or six non-resupinate flowers, on compressed peduncle. Sepals and petals thick, pale yellow-brown on face, heavily overlaid with dark red-maroon except at apical margins, backs green with similar but lighter flush, apices thickened and keeled as also midvein of lip. Lip three-lobed, lateral lobes larger and wider than midlobe, thick and rigid, off-white with purple flush on back. Face of lip covered with fine purple veining on lateral lobes contrasting with clearly delineated pure white margin. Callus near base purple covered with fine white hairs. Column cream. Anther yellow.

HABITAT AND DISTRIBUTION

Colombia and Venezuela. Found at 1600–2000 meters in the cloud forest.

MEASUREMENTS

Pseudobulbs 5–7 cm long, 2 cm wide
Leaves 15 cm long, 2 cm wide
Inflorescence 15 cm long
Sepals 20 mm long, 7 mm wide
Petals 17 mm long, 7 mm wide
Lip 13 mm long, 15 mm wide
Column 1 mm wide

Anacheilium mejia

FIGURE 1-30, PLATE 36

Anacheilium mejia Withner & Harding, *sp. nov.*
> *Pseudobulbi 25 × 1.5 cm plus minusve compressi in rhizomate ad intervalla 8 cm separati. Folia tres, 28 × 2.5 cm. Inflorescentia ex vagina emergens. Pedicellus ovariumque 25 mm. Flores luteovirides, sepals petalisque lanceolatis 11 × 3 mm, labio convexo 7 × 7 mm luteo purpureo-guttato, callo bicarinato in medio sulcato apice purpureo, columna 5 mm longa unguiculata ad labium e basi plus minusve 2 mm connata. Species affinis Anacheilium sceptro.* Type: Colombia, Hort. Mejia de Moreno (holotype: US)

DERIVATION OF NAME

Honors Esperanza Mejia de Moreno of Colombia, who exhibited the huge, well-grown plant that drew our attention.

DESCRIPTION

Pseudobulbs spindle-shaped, slightly compressed. Leaves three. Inflorescence emerges from mature growths with a spathe. Flowers many (to twenty), yellow to yellow-green. Petals rolled in at sides. Lip yellow with dark maroon venation, which extends from end of callus to near apex, leaving margins of lip free of maroon coloration. Lip three-lobed, though lateral lobes are small. Callus with two lobes and a central sulcus.

HABITAT AND DISTRIBUTION

Colombia.

FLOWERING TIME

August and September.

MEASUREMENTS

Pseudobulbs 25 cm long, 1.5 cm wide
Leaves 28 cm long, 2.5 cm wide
Inflorescence 15–30 cm long
Spathe 2.5 cm long
Sepals 11–12 mm long, 2–3 mm wide
Petals 11 mm long, 3 mm wide
Lip 7 mm long, 7 mm wide
Column 5 mm long, 1.5 mm wide

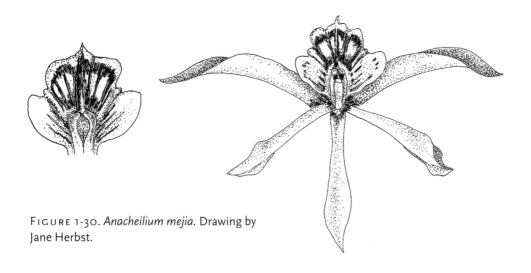

FIGURE 1-30. *Anacheilium mejia.* Drawing by Jane Herbst.

Anacheilium moojenii

Anacheilium moojenii (Pabst) Pabst, Moutinho & A. V. Pinto. 1981. *Bradea* 3: 23. Basionym: *Epidendrum moojenii* Pabst. 1956. *Orquídea* 17: 204. Type: Brazil, Bahia, Chapada Diamantina, between Lençois and Palmeiras, Municipality of Castro Alves, 300 meters, Collector *J. Moojen de Oliveira s.n.*, flowered in plantarium Hort. Fluminensis November 1955. (holotype: RB 93537).

SYNONYMS
Encyclia moojenii (Pabst) Pabst. 1967. *Orquídea* 29 (6): 276.
Hormidium moojenii (Pabst) Brieger. 1977. Schlechter's *Die Orchideen*, ed. 3, p. 571.
Prosthechea moojenii (Pabst) W. E. Higgins. 1997. *Phytologia* 82 (5): 381.

DERIVATION OF NAME
Honors a collector, J. Moojen de Oliveria

DESCRIPTION
A rock- and cliff-dwelling, moderately robust plant with a thick fleshy rhizome that is elongate and sheathed. Pseudobulbs robust ovoid, slightly compressed, smooth with two paired leathery sheaths. Leaves two or three, leathery, intensely green. Flowers seven to ten, spreading moderately. Flowers uniformly white-yellow. Sepals pointed, eleven-nerved, petals like sepals. Lip fleshy, outline rhomboid with the apex being the narrower end, lip fused with column basally. Callus linear, one half a longitudinal groove, the other half a hollowed out curve, apex ending abruptly with three teeth. Column straight, three-angled, broadened at apex. Ovary three-angled. Anther yellow.

HABITAT AND DISTRIBUTION
Brazil (Bahia and Minas Gerais). Found on rocks with humus, at 300 meters in elevation.

COMMENT
This species is reportedly spider pollinated, the spider web being the same color as the flowers (Pabst 1956, *Orquídea* 17: 205).

MEASUREMENTS
Pseudobulbs 8 cm long, 2–3 cm wide
Leaves 18–25 cm long, 2.5–3 cm wide
Inflorescence 3 cm long
Spathe small

Sepals 20–22 mm long, 6–7 mm wide
Petals 18 mm long, 5 mm wide
Lip 15 mm long, 10 mm wide
Column 10 mm long

Anacheilium neurosum

Anacheilium neurosum (Ames) Withner & Harding, *comb. nov.* Basionym: *Epidendrum neurosum* Ames. 1922. *Schedulae Orchidianae* 1: 17.

SYNONYMS
Encyclia neurosa (Ames) Dressler & Pollard. 1971. *Phytologia* 21: 437.
Prosthechea neurosa (Ames) W. E. Higgins. 1997. *Phytologia* 82 (5): 381.

DERIVATION OF NAME
Latin *neuro,* "nerved," referring to the lip veins.

DESCRIPTION
Pseudobulbs spaced 2–4 cm apart, narrowly spindle-shaped, shortly stalked, somewhat flattened. Leaves two. Inflorescence short. Flowers two, cream color, lip with several violet lines, especially near base. Lip cordate at base, forming true heart shape. Column without toothlike lateral wings on the sides to distinguish it from *Anacheilium abbreviata* column, which is wider in the middle, winglike.

HABITAT AND DISTRIBUTION
Costa Rica, Guatemala, Honduras, and México. Found at 900–1100 meters in wet tropical evergreen forest.

FLOWERING TIME
October to December.

MEASUREMENTS
Pseudobulbs 3.5–9 cm long, 0.6–1.4 cm wide
Leaves 8–19 cm long, 0.6–1.5 cm wide
Inflorescence 2–3 cm long
Spathe 4 cm long
Sepals 16–26 mm long, 2.8–4 mm wide
Petals 15–21 mm long, 3.2–5 mm wide
Lip 14–18.5 mm long, 5.5–8.5 mm wide
Column 5–5.5 mm long

Anacheilium pamplonense

Anacheilium pamplonense (Reichenbach f.) Withner & Harding, *comb. nov.*
Basionym: *Epidendrum pamplonense* Reichenbach f. 1849. *Linnaea* 22: 837.

SYNONYMS

Encyclia pamplonensis (Reichenbach f.) Carnevali. 1984. *Phytologia* 55 (5): 288.
Prosthechea pamplonensis (Reichenbach f.) W. E. Higgins. 1997. *Phytologia* 82 (5): 379.

DERIVATION OF NAME

After Pamplona, Venezuela, the place where the plant was originally collected.

DESCRIPTION

An epiphyte, terrestrial, or lithophyte. Pseudobulbs tall, slender. Leaves two. Flowers up to fourteen on an upright raceme, fleshy, rigid. Tepals yellow and white flecked with purple, lip pink. Lip squarish, unlobed or obscurely three-lobed with an apical notch. Callus not extending past column, with two ridges and a small sulcus in between. Column yellow and flecked with purple. Ovary three-winged.

HABITAT AND DISTRIBUTION

Venezuela (Pamplona). Found at 1800 meters in elevation.

FLOWERING TIME

January.

COMMENT

This species, along with *Anacheilium vespa*, *A. crassilabium*, and *A. tigrinum*, has been confused in the past, with some authors lumping them all together. We present here a small key to differentiate this group:

    Lip pointed . . . . *Anacheilium crassilabium*, *A. vespa*
    Lip rounded . . . . *Anacheilium tigrinum*
    Lip notched, squarish . . . . *Anacheilium pamplonense*

MEASUREMENTS

Pseudobulbs 40 cm long, 1 cm wide
Leaves 20 cm long, 7 cm wide
Inflorescence 20 cm long
Spathe 1 cm long
Sepals 20 mm long, 10 mm wide
Petals 18 mm long, 11 mm wide
Lip 1 mm long, 1 mm wide
Column 5 mm long

Anacheilium papilio
FIGURE 1-31, PLATE 37

Anacheilium papilio (Vellozo) Pabst, Moutinho & A. V. Pinto. 1981. *Bradea* 3: 23. Basionym: *Epidendrum papilio* Vellozo. 1831. *Florae Fluminensis* 9: t. 28.

SYNONYMS
Encyclia papilio (Vellozo) Pabst. 1967. *Orquídea* 29 (6): 276.
Prosthechea papilio (Vellozo) W. E. Higgins. 1997. *Phytologia* 82 (5):3 81.
Hormidium papilio (Vellozo) Brieger ex The International Plant Names Index.

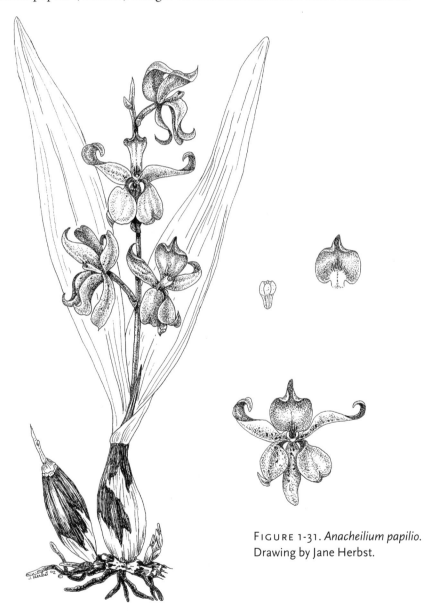

FIGURE 1-31. *Anacheilium papilio.* Drawing by Jane Herbst.

DERIVATION OF NAME
Latin *papilio,* "butterfly," referring to the flower's appearance.

DESCRIPTION
Pseudobulbs narrowed, conical-oblong, clustered. Leaves two. Inflorescence many-flowered. Sepals longer and more narrow than petals, tips curled. Lip unlobed, adnate for basal 2–3 mm, obcordate, widest at midsection. Callus composed of two keels, which diverge midway and converge at apex of callus.

HABITAT AND DISTRIBUTION
Brazil (Espírito Santo, Minas Gerais, Rio de Janeiro, Rio Grande do Sul, and Santa Catalina). Found in open forest.

COMMENT
Figure 1-31 does not show the keels on the lip, as the artist worked from a picture and was unable to see the callus structure in the photo. Figure 1-3 shows the keels; this drawing was taken from Pabst and Dungs (1977).

MEASUREMENTS (taken from drawings)
Pseudobulbs 12 cm long, 4 cm wide
Leaves 24 cm long, 5 cm wide
Inflorescence 32 cm long
Sepals 22 mm long, 4 mm wide
Petals 20 mm long, 8 mm wide
Lip 17 mm long, 8 mm wide

Anacheilium radiatum
FIGURE 1-32, PLATE 38

Anacheilium radiatum (Lindley) Pabst, Moutinho & A. V. Pinto. 1981. *Bradea* 3: 23. Basionym: *Epidendrum radiatum* Lindley. 1841. *Edward's Botanical Register,* misc. 58.

SYNONYMS
Epidendrum marginatum Link, Klotzsch & Otto. 1842. *Ic. Pl. Rar. Hort. Berol.* 90: t. 6.
Encyclia radiata (Lindley) Dressler. 1961. *Brittonia* 13 (3): 265.
Prosthechea radiata (Lindley) W. E. Higgins. 1997. *Phytologia* 82 (5): 381.

DERIVATION OF NAME
Latin *radiatum,* "in a radiating manner," referring to the purple veins of the lip.

DESCRIPTION

A stout plant 12–40 cm tall. Pseudobulbs stalked (stipitate), ovoid to spindle-shaped, compressed, strongly ribbed, and usually grooved. Leaves two to four. Flowers seven to ten, non-resupinate, showy, pale greenish white or yellowish green, lip striate with purple. Lip unlobed, cupped, wider than long, apex rounded, some varieties with a point, others with a central notch with a small triangular

FIGURE 1-32. *Anacheilium radiatum*. Drawing by Jane Herbst.

point at midpoint. Callus pubescent on distal half. Midtooth merging into broad ligule (fimbriate). Capsule sharply triangular in cross section. Flowers strongly and pleasantly fragrant.

HABITAT AND DISTRIBUTION
México to Venezuela. Found on trees and rocks in dense or open forests up to 1200 meters in elevation.

FLOWERING TIME
April to July.

CULTURE
Culture for this species is intermediate to warm temperatures, constant moisture to root area in spring and summer, less so in the winter. High light is preferred but plants will tolerate bright diffuse light.

COMMENT
This is another species that amazingly does not have a long list of synonyms.

MEASUREMENTS
Pseudobulbs 7–11 cm long, 1.2–3 cm wide
Leaves 14–35 cm long, 1.4–3 cm wide
Inflorescence 7–24 cm long
Sepals 15–20 mm long, 5–7 mm wide
Petals 15–20 mm long, 8–11 mm wide
Lip 10–13 mm long, 13–20 mm wide
Column 8–10 mm long

Anacheilium regnellianum

Anacheilium regnellianum (Hoehne & Schlechter) Withner & Harding, *comb. nov.* Basionym: *Epidendrum regnellianum* Hoehne & Schlechter. 1926. *Arquivos de botanica do estado de São Paulo*, 1: 243. Type: Brazil, Minas Gerais: Pedro Branca, Caldas, collected by F. C. Hoehne, flowered in cultivation by Oswaldo Cruz 14 August 1920, *F. C. Hoehne 4320* (holotype: SP).

SYNONYMS
Encyclia regnelliana (Hoehne & Schlechter) Pabst. 1967. *Orquídea* 29 (6): 276.
Prosthechea regnelliana (Hoehne & Schlechter) W. E. Higgins. 1997. *Phytologia* 82 (5): 381.

DERIVATION OF NAME
Honors Dr. Anders Regnell.

DESCRIPTION
An epiphyte with a short, moderately thick rhizome. Pseudobulbs oblong, spindle-shaped. Leaves one to three, thick. Flowers five to nine. Sepals flesh pink colored, with purple-red spots near base, petals whitish, spotted irregularly. Lip adnate to column basally, lamina distinctly three-lobed, callus square. Column three-toothed, white.

HABITAT AND DISTRIBUTION
Brazil (Minas Gerais).

COMMENT
We have seen a drawing of this plant, but the drawing was not well reproduced and we could not get enough detail to get a redrawing of this flower. From what we can see, the lip is small and not a prominent or showy portion of the flower. It is hard to say for sure that the column teeth are the same length, which makes the validity of placing it in *Anacheilium* as opposed to another genus (*Prosthechea*) uncertain. The text along with the type description mentions that it may be similar to *Epidendrum brachychilum* (synonym *Anacheilium brachychilum*). In that case, you will have trouble getting to this species' page with the key. It also says it is like *Epidendrum inversum* (synonym *Anacheilium bulbosum*), and in that case, you should get here with the key.

MEASUREMENTS
Pseudobulbs 4–6 cm long, 1–2 cm wide
Leaves 18–20 cm long, 1.8–2.5 cm wide
Inflorescence 6–11 cm long
Sepals 20 mm long
Petals 16 mm long, 8 mm wide
Lip 8 mm long, 9 mm wide
Column 8 mm long

Anacheilium santanderense
FIGURE 1-33, PLATE 39

Anacheilium santanderense Hunt, Withner & Harding, *sp. nov.*
Pseudobulbi 13 × 1.5 cm plus minusve compressi in rhizomate ad intervalla 2 cm separati. Folia duo 22.4 × 2.9 cm. Inflorescentia ex vagina emergens.

Pedicellus ovariumque 13 mm. Flores luteovirides, sepals petalisque lanceolatis 8 × 3 mm, labio convexo 8 × 5 mm luteo purpureo-guttato, callo bicarinato in medio sulcato apice purpureo, columna 4.5 mm longa unguiculata ad labium e basi plus minusve 1 mm connata. Species floribus minimus inter species affines Anacheilium calamario vel Anacheilium carrio columnas claviformes praebens. Type: Colombia. Near San Vincente de Chucur, Department of Santander, 1200–600 meters, flowered in cultivation by David Hunt (holotype: US).

DERIVATION OF NAME
After Santander, the plant's place of origin.

DESCRIPTION
An epiphyte. Pseudobulbs slightly compressed, separated along the rhizome at 2-cm-long intervals. Leaves two. The inflorescence emerges from a sheath. Pedicel and ovary 13 mm long. Flowers yellow-green. Lip yellow with maroon spots. Callus two keels with central sulcus, maroon at the apex of the callus. Column clawed and basally adnate for 1 mm with the lip.

HABITAT AND DISTRIBUTION
Colombia.

FLOWERING TIME
Bloomed August 2001 in cultivation in Texas.

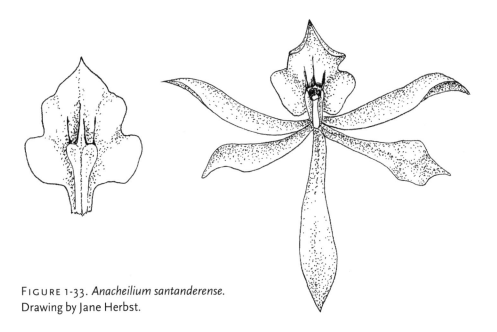

FIGURE 1-33. *Anacheilium santanderense.*
Drawing by Jane Herbst.

MEASUREMENTS

Pseudobulbs 13–25 cm long, 1.5 cm wide
Leaves 22.4 cm long, 2.9 cm wide
Inflorescence 55 cm long
Sepals 8–14 mm long, 3–4.5 mm wide
Petals 8–14 mm long, 3–5 mm wide
Lip 8–10 mm long, 5–12 mm wide
Column 4.5 mm long

Anacheilium sceptrum
FIGURE 1-34, PLATE 40

Anacheilium sceptrum (Lindley) Hágsater. 1990. *Native Colombian Orchids* 1: 28.
Basionym: *Epidendrum sceptrum* Lindley. 1846. *Orchidaceae Lindenianae* 8.

SYNONYMS

Epidendrum macrothyrsoides Reichenbach f. 1877. *Linnaea* 41: 113.
Epidendrum sphenoglossum F. C. Lehmann & Kraenzlin. 1899. *Engl. Bot. Jahrb. Syst.* 26: 460. Type: Ecuador. In rocks by the rivers Luis and Calera near Zaruma, 800–1200 meters, December 1890, *F. C. Lehmann 8178* (holotype: K; photo: RPSC).
Encyclia sceptra (Lindley) Carnevali & I. Ramírez. 1986. *Ernestia* 36: 9.
Prosthechea sceptra (Lindley) W. E. Higgins. 1997. *Phytologia* 82 (5): 381.

DERIVATION OF NAME

Sceptrum, "staff" or "wand," referring to the inflorescence.

DESCRIPTION

Pseudobulbs at intervals of 3–6 cm, compressed. Raceme, spikelike, 30–60 cm long. Flowers numerous. Sepals and petals golden yellow spotted with purple. Lip bright purple, white at base. Lip adnate to column basally, very weakly trilobed, rounded spear-shaped. Column fleshy, three-toothed, pale brown, the apex a prominent fleshy crest below which a narrow lamina protrudes to lie across the anther.

HABITAT AND DISTRIBUTION

Colombia, Ecuador, and Venezuela. Found at 150–1650 meters in elevation.

FLOWERING TIME

August and occasionally in January.

CULTURE

Culture for this plant is intermediate to warm temperatures with strong light and constant moisture, more so in the warm months.

COMMENT

We saw several plants in 2002 sold as *Anacheilium sceptrum*, but they turned out to be one of the new species we describe here, *A. vita*. *Anacheilium sceptrum* has little or no callus; this new species has a definite callus.

FIGURE 1-34. *Anacheilium sceptrum*. Drawing by Jane Herbst.

MEASUREMENTS

Pseudobulbs 25 cm long, 1 cm wide
Leaves 30 cm long, 3 cm wide
Inflorescence 55 cm long
Sepals 14 mm long, 5 mm wide
Petals 14 mm long, 5 mm wide
Lip 10 mm long, 12 mm wide
Column 2 mm long

Anacheilium sessiliflorum

Anacheilium sessiliflorum (Edwall) Withner & Harding, *comb. nov.* Basionym: *Epidendrum sessiliflorum* Edwall. 1903. *Rev. Centr. Sci. Camp* 4 extr: 3, tab. 4.

SYNONYMS

Encyclia sessiliflora (Edwall) Pabst. 1967. *Orquídea* 29 (6): 277.
Hormidium sessiliflorum (Edwall) Pabst, Moutinho & A. V. Pinto. 1981. *Bradea* 3: 23.
Prosthechea sessiliflora (Edwall) W. E. Higgins. 1997. *Phytologia* 82 (5): 381.

DERIVATION OF NAME

Sessili, "sessile," and *florum,* "flowers," referring to the short inflorescence.

DESCRIPTION

Pseudobulbs clustered, spindle-shaped, slightly compressed, apex tapered, sheath covering lower portion of pseudobulb. Leaves one or two. Flowers three. Sepals and petals fleshy, subequal, concave, pale yellow. Lip attached to column adnate basally, unlobed, apex rounded, strongly concave, yellow. Capsule three-edged, green.

HABITAT AND DISTRIBUTION

Brazil (São Paulo and Minas Gerais).

FLOWERING TIME

January.

COMMENT

Eric Christenson (pers. comm.) notes that the drawing of the type shows fruit on the inflorescence, indicating that it may be cleistogamous, and he notes that it blooms on the still forming new growth. He feels it is probably close to *Anacheilium calamarium*.

MEASUREMENTS
Pseudobulbs 7 cm long, 2.5 cm wide
Leaves 16 cm long, 2.5 cm wide
Inflorescence sessile
Spathe 1–1.5 cm long
Sepals 10 mm long, 7 mm wide
Petals 8 mm long, 4 mm wide
Lip 5 mm long, 4 mm wide
Column 5 mm long

Anacheilium simum
FIGURE 1-35, PLATE 41

Anacheilium simum (Dressler) Withner & Harding, *comb. nov.* Basionym: *Encyclia sima* Dressler. 1969. *Orquideologia* 4 (2): 91. Type: Panama, Province of Panamá, Cerro Jefe, 950 meters, 12 November 1967, *Robert L. Dressler 3160* (holotype: US).

SYNONYMS
Hormidium simum (Dressler) Brieger. 1977. Schlechter's *Die Orchideen*, ed. 3, p. 569, as *Hormidium sima*.
Prosthechea sima (Dressler) W. E. Higgins. 1997. *Phytologia* 82 (5): 381.

DERIVATION OF NAME
Latin *sima*, "pug-nosed," referring to the form of the column.

DESCRIPTION
Pseudobulbs elongate. Leaf one. Inflorescence a few-flowered erect raceme. Sepals and petals pointed, lip acuminate (acutely pointed), with moderately thick callus, base of lip fused to column. Petals rose pink on the back, white-cream on the front. Lip cream white with pink-violet veins. Flowers very fragrant.

HABITAT AND DISTRIBUTION
Panama and Colombia. Found at 950 meters, in humid forest.

CULTURE
Patricia has had this plant for years, after obtaining it from Bill Leonard, who imported it years ago. It is one of her favorite plants. The plant is in a basket, where it gets high diffuse light and year-round moisture in intermediate to cool conditions.

MEASUREMENTS

Pseudobulbs 8–20 cm long, 1–1.5 cm wide
Leaves 2–25 cm long, 2.5–4 cm wide
Inflorescence 4.5–5.5 cm long
Spathe 0.5–0.7 cm long
Sepals 28–37 mm long, 8 mm wide
Petals 26–31 mm long, 10 mm wide
Lip 22–24 mm long, 8–10 mm wide
Column 7–8 mm long, 6–7 mm wide

FIGURE 1-35. *Anacheilium simum.*
Drawing by Jane Herbst.

Anacheilium spondiadum
PLATE 42

Anacheilium spondiadum (Reichenbach f.) Nir ex Withner & Harding, *comb. nov.* Basionym: *Epidendrum spondiadum* Reichenbach f. 1852. *Botanische Zeitung (Berlin)* 10: 731.

SYNONYMS
Epidendrum playcardium Schlechter. 1922. *Repertorium Specierum Novarum Regni Vegetabilis, Beihefte* 17: 36.
Encyclia spondiada (Reichenbach f.) Dressler. 1971. *Phytologia* 21 (7):441.
Hormidium spondiadum (Reichenbach f.) Brieger. 1977. Schlechter's *Die Orchideen,* ed. 3, 569.
Prosthechea spondiada (Reichenbach f.) W. E. Higgins. 1997. *Phytologia* 82 (5): 381.
Anacheilium spondiadum (Reichenbach f.) Nir. 2000. *Orchidaceae Antillanae,* p. 319. *nom. nud.*

DERIVATION OF NAME
Greek *spondias,* "a kind of plum tree," referring to the flower color.

DESCRIPTION
An epiphyte to 35 cm tall. Pseudobulbs cylindric to narrowly ovoid. Leaf one. Flowers few (three or four), non-resupinate. Sepals and petals greenish white to white with red overlay distally, sometimes the red almost covers entire tepals, the lip is rose to plum purple. Lip unlobed, elliptic-suborbicular, pointed, cupped. Callus basal. Column three-toothed, irregularly divided. Ovary three-winged.

HABITAT AND DISTRIBUTION
Costa Rica, Jamaica, and Panama. Found at 1000–2100 meters in elevation.

FLOWERING TIME
November to March.

MEASUREMENTS
Pseudobulbs 3–12 cm long, 0.6–1.3 cm wide
Leaves 10–28 cm long, 2.5–5 cm wide
Inflorescence shorter than leaves
Spathe 2.5 cm long
Sepals 13–32 mm long, 3–6 mm wide
Petals 12–28 mm long, 3.5–9 mm wide

Lip 12–20 mm long, 8–13 mm wide
Column 5–8 mm long, 3.5 mm wide

Anacheilium suzanense
FIGURE 1-36

Anacheilium suzanense (Hoehne) Pabst, Moutinho & A. V. Pinto. 1981. *Bradea* 3: 23. Basionym: *Epidendrum suzanense* Hoehne. 1938. *Arquivos de botanica do estado de São Paulo*, 1, as *E. susanense*.

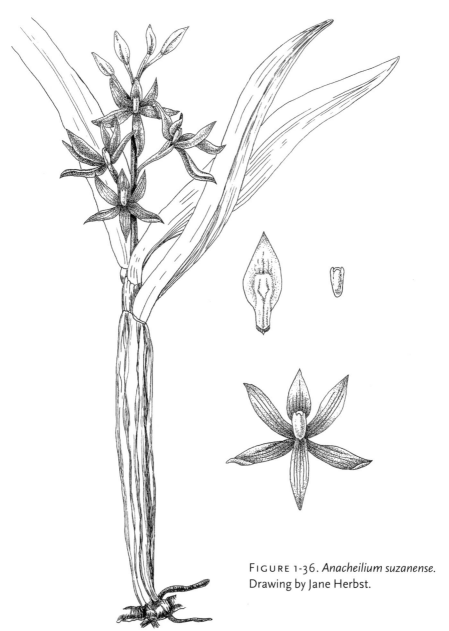

FIGURE 1-36. *Anacheilium suzanense*. Drawing by Jane Herbst.

SYNONYMS

Encyclia suzanensis (Hoehne) Pabst. 1967. *Orquídea* 29 (6): 276.
Prosthechea suzanensis (Hoehne) W. E. Higgins. 1997. *Phytologia* 82 (5): 381.

DERIVATION OF NAME

After Suzano, a municipality of Mogí das Cruzes, São Paulo. We have seen this species name spelled with a *z* or an *s*. The original description uses *s*, but the drawing name associated with the description uses *z*. Because Brazilians spell the city name with *z*, the correct spelling for the scientific name is *Anacheilium suzanense*.

DESCRIPTION

An epiphyte. Pseudobulbs spaced 1–2 cm apart on a robust rhizome. Leaves three. Racemes to 20 cm long. Flowers ten to fifteen. Flowers yellow-white, disc of lip violet. Lip ovoid coming to point. Callus white with purple at base to apex. Wings running down the column.

HABITAT AND DISTRIBUTION

Brazil (Minas Gerais and São Paulo).

COMMENT

We have no experience with this plant. We were considering including it in the *papilio- widgrenii* group but as we cannot find enough evidence to determine whether this is correct, we leave it separate. Hoehne in his original description suggests it may be a natural hybrid between *Anacheilium widgrenii* and *A. vespa* and that its habitat is the same as that of *A. vespa*.

MEASUREMENTS

Pseudobulbs 15–23 cm long, 2 cm wide
Leaves 20 cm long, 2.5–3.5 cm wide
Inflorescence 20 cm long
Sepals 20 mm long, 4.5 mm wide
Petals 20 mm long, 4 mm wide
Lip 15 mm long
Column 10 mm long

Anacheilium tigrinum

PLATE 43

Anacheilium tigrinum (Linden ex Lindley) Wither & Harding, *comb. nov.* Basionym: *Epidendrum tigrinum* Linden ex Lindley. 1846. *Orch. Linden.* 9.

SYNONYMS

Aulizeum tigrinum Lindley ex Stein. 1892. *Orchideenbuch* 240.

Hormidium tigrinum (Lindley) Brieger. 1977. Schlechter's *Die Orchideen*, ed. 3, p. 572.

Encyclia tigrina (Linden ex Lindley) Carnevali & Ramirez. 1986. *Ernestia* 36: 9.

Prosthechea tigrina (Linden ex Lindley) W. E. Higgins. 1997. *Phytologia* 82 (5): 381.

DERIVATION OF NAME

Latin *tigrinus*, "spotted like a jaguar," referring to the flowers.

DESCRIPTION

A terrestrial or an epiphyte. Pseudobulbs elongate, very tall and slender, lightly compressed. Flowers on upright raceme, 3 cm in diameter. Ovary slightly warty. Tepals thick and fleshy, yellow with many purple spots. Lip very rounded, yellow with fainter purple markings. Column yellow, callus hairy, with biconcave corrugations (raised ridges) extending to lip apex.

HABITAT AND DISTRIBUTION

Brazil, Ecuador, Peru, Venezuela, and the Guianas. Found at 2100–2700 meters in dense forests.

FLOWERING TIME

July to February.

CULTURE

In an article in *The Orchid Digest* in 1980 (36 [2]: 53), Dunsterville reports that *Anacheilium tigrinum* is more difficult to grow than its *A. crassilabium* relatives.

COMMENT

Plate 43 seems to show the flowers being lip down; because the camera is looking down on the flowers from above, the photo gives the illusion that the lip is down. The flowers are non-resupinate.

MEASUREMENTS (taken from drawing of basionym)

Pseudobulbs 30 cm long, 1.2 cm wide
Leaves 20–25 cm long, 3.7 cm wide
Inflorescence 20–30 cm long
Spathe 2.5 cm long
Sepals 20 mm long, 10 mm wide
Petals 20 mm long, 11 mm wide
Lip 13 mm long, 10 mm wide
Column 8–10 mm long

Anacheilium trulla

FIGURE 1-37, PLATE 44

Anacheilium trulla (Reichenbach f.) Withner & Harding, *comb. nov.* Basionym: *Epidendrum trulla* Reichenbach f. 1856. *Bonplandia* 4: 214. Type: México, *Herb. Pavon s.n.* (holotype: G; fragment: W).

SYNONYMS

Epidendrum langlassei Schlechter. 1918. *Beih. Bot. Centralbl.* 36 (2): 404.
Encyclia trulla (Reichenbach f.) Christenson. 1996. *Lindleyana* 11 (4): 222.
Prosthechea trulla (Reichenbach f.) W. E. Higgins. 1997. *Phytologia* 82 (5): 381.

Not *Epidendrum lancifolium* Pavón ex Lindley, 1831, *Gen. Sp. Orch. Pl.* 98. This is
    Anacheilium cochleatum per Christenson 1996. As this is an invalid name, the
    next available name is *Epidendrum trulla* Reichenbach f. published in 1856.

DERIVATION OF NAME

Latin *trulla*, "trowel-shaped," referring to the lip.

DESCRIPTION

An epiphyte, sometimes a lithophyte. Pseudobulbs loosely clustered, spaced 1.5 cm apart on a rhizome, spindle-shaped or ellipsoid, flattened, with distinct basal stalk. Leaves two. Inflorescence a raceme of four to ten flowers. Sepals and petals cream to greenish to yellowish white. Lip cupped, green heavily lined with purple-violet on basal two thirds, apex heavily stained yellow-green within, purplish on reverse. There are two or three more or less developed ridges or elongated swellings at the base of the lip, lip is variably five-angled.

HABITAT AND DISTRIBUTION

México. Found at 300–1500 meters, in oak or pine-oak forest, sometimes on rocks.

CULTURE

Culture is the same as for *Anacheilium cochleatum*, though requiring a bit more light than *Anacheilium cochleatum*, with intermediate to warm temperatures and year-round watering. Plant is not everblooming, unlike *A. cochleatum*.

COMMENT

This species is closely related to *Anacheilium cochleatum*, but its shorter sepals and petals and paler lip distinguish it. On first glance this plant and flower look like a miniature version of *A. cochleatum*, but the plant is smaller and more manageable and the stem is held nicely just above the foliage, the blooms looking like a smaller octopus.

MEASUREMENTS
Pseudobulbs 7–14 cm long, 2–5 cm wide
Leaves 12–27 cm long, 1.8–4 cm wide
Inflorescence 5–12 cm long
Sepals 15–22 mm long, 4–8 mm wide
Petals 13–18 mm long, 4.5–8 mm wide
Lip 9–12 mm long, 9–17 mm wide
Column 7–7.5 mm long

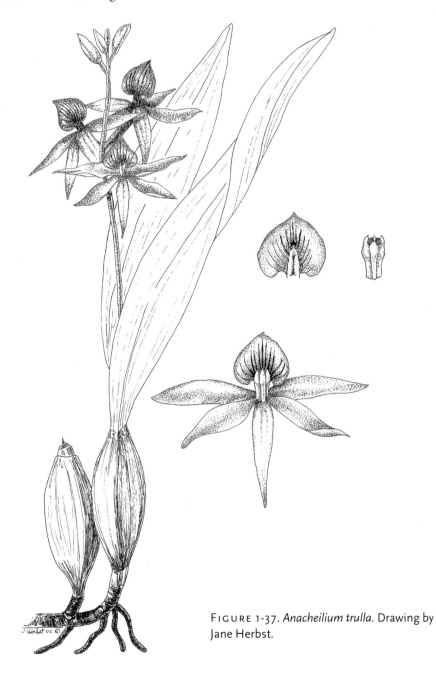

FIGURE 1-37. *Anacheilium trulla*. Drawing by Jane Herbst.

Anacheilium vagans
FIGURE 1-38, PLATE 45

Anacheilium vagans (Ames) Withner & Harding, *comb. nov.* Basionym: *Epidendrum vagans* Ames. 1923. *Schedulae Orchidianae* 6: 76. Type: Costa Rica: southern flanks of Volcano Irazú, 1500–2100 meters, *H. Lankester 461* (holotype: AMES, no. 26948).

SYNONYMS
Encyclia vagans (Ames) Dressler & Pollard. 1971. *Phytologia* 21 (7): 438.
Prosthechea vagans (Ames) W. E. Higgins. 1997. *Phytologia* 82 (5): 381.

DERIVATION OF NAME
Latin *vagus*, "in no particular direction," referring to the plant's growth pattern.

DESCRIPTION
Rhizome elongated, creeping, 5–6 mm in diameter. Pseudobulbs 5–6 cm apart, spindle-shaped, concealed when young by sheaths. Leaves three or four. Inflorescence a loose raceme of four flowers, non-resupinate, white or cream, the sepals and petals with a short median band of purple, the lip with about twelve purple stripes. Sepals with deep keel along middle of outer surface. Lip elliptic-ovate, tapering to point and slightly keeled on the under side near tip, concave. Callus velutinous, pubescence lacking in a median groove. Capsule three-winged.

HABITAT AND DISTRIBUTION
Costa Rica, El Salvador, Guatemala, México, and Peru. Found at 1400–1950 meters, in wet pine-oak forest.

FLOWERING TIME
March to June.

CULTURE
We have no cultural information on this plant. Keeping it in a pot, however, would be difficult due to its rambling habit.

COMMENT
Three plants—*Anacheilium abbreviatum* (Plate 1), *A. alagoense* (Plate 3), and *A. vagans* (Plate 45)—appear very similar, having the red line down the tepals. Compare the description carefully to make certain of your identification.

This species differs from *Anacheilium radiatum* in the outline of lip and keel on the lateral sepals. It is also similar to *A. ionophlebium* in habitat, but has smaller leaves, and pseudobulbs well separated on a creeping rhizome.

MEASUREMENTS
Pseudobulbs 5–8 cm long, 1.2 cm wide
Leaves 8–15 cm long, 1.6–2.2 cm wide
Inflorescence 8 cm long
Sepals 18–20 mm long, 4 mm wide
Petals 16 mm long, 4–5 mm wide
Lip 14 mm long, 7–9 mm wide
Column 5.5–7 mm long

FIGURE 1-38. *Anacheilium vagans.* Drawing by Jane Herbst.

Anacheilium vasquezii
FIGURE 1-39, PLATE 46

Anacheilium vasquezii (Christenson) Withner & Harding, *comb. nov.* Basionym: *Prosthechea vasquezii* Christenson. 2003. *Richardiana* 3(3): 117. Type: Bolivia, Department Cochabamba, Province Chapare, Kilometer 100, Cochabamba-Villa Tunari, 1880 meters, *R. Vásquez 27* (holotype: Herb. Vásquezianum).

DERIVATION OF NAME
Honors Roberto Vásquez Ch., discoverer and illustrator of this species, who has made many contributions to Bolivian orchidology.

DESCRIPTION
An epiphyte with a long, trailing horizontal rhizome. Pseudobulbs spindle-shaped, separated by up to 10 cm along rhizome. Leaves two or three. Inflorescence erect, laxly flowered. Flowers non-resupinate, about 2.8 cm wide. Sepals and petals green at the base with yellow margins and apex, marked with transverse reddish maroon bars except at the unmarked tips. Lip dark red with white margins, three-lobed, the lateral lobes suborbicular, obtusely rounded without a distinct apex, the midlobe elliptic, rounded with a minute apicule, midlobe smaller than the lateral lobes. Callus quadrate, sulcate.

HABITAT AND DISTRIBUTION
Bolivia. Found at 1880 meters in elevation.

COMMENT
Plate 46 shows a plant from Peru, which would be a range extension for this species. The angle of the photograph does not let us get a close look at the lip to make sure, and it is more truly identified by excluding others.

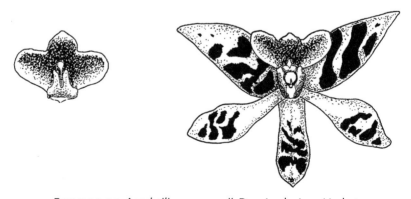

FIGURE 1-39. *Anacheilium vasquezii*. Drawing by Jane Herbst.

Anacheilium vasquezii is similar to *A. hajekii* but differs by being epiphytic, having long rhizome segments, and being more laxly flowered, with rounded lateral lip lobes and a different callus.

Anacheilium vasquezii is also similar to *A. hartwegii* but differs by having a midlobe smaller than the lateral lobes, and less broad leaves.

MEASUREMENTS

Pseudobulbs 20 cm long
Leaves 22 cm long, 2.5 cm wide
Inflorescence 20 cm long
Spathe 7 cm long
Sepals 13 mm long, 6 mm wide
Petals 11 mm long, 5 mm wide
Lip 8 mm long, 6 mm wide
Column 6 mm long

Anacheilium venezuelanum
PLATE 47

Anacheilium venezuelanum (Schlechter) Withner & Harding, *comb. nov.* Basionym: *Epidendrum venezuelanum* Schlechter. 1919. *Repertorium Specierum Novarum Regni Vegetabilis, Beihefte* 6: 39. Type: Venezuela, Federal District, Caracas, June 1905, *Hort. K. W. John s.n.* (holotype: B, destroyed).

SYNONYMS
Encyclia venezuelana (Schlechter) Dressler. 1971. *Phytologia* 21: 441.
Prosthechea venezuelana (Schlechter) W. E. Higgins. 1997. *Phytologia* 82 (5): 381.

DESCRIPTION
An epiphyte with a short rhizome. Pseudobulbs oblong-ovoid, flattened, ribbed. Leaf one. Inflorescence an erect raceme on mature growth. Flowers five to seven, non-resupinate, greenish white, marked with linear red-brown spots on the segments. Callus yellow. Lip elliptical with red-brown stripes on lip, narrowing to point, keel-like extension down center of lip. Column winged. Pedicel and ovary three-angled, 8 mm long.

HABITAT AND DISTRIBUTION
Colombia, Ecuador, and Venezuela. In wet montane forest.

FLOWERING TIME
May to August.

MEASUREMENTS
Pseudobulbs 6–7 cm long, 0.6–0.8 cm wide
Leaves 7–13 cm long, 1.3–1.8 cm wide
Peduncle 2.5 cm long
Spathe 1.5 cm long
Sepals 14 mm long
Petals 13 mm long
Lip 12 mm long
Column 7 mm long

Anacheilium vespa
FIGURE 1-40, PLATE 48

Anacheilium vespa (Vellozo) Pabst, Moutinho & A. V. Pinto. 1981. *Bradea* 3: 23. Basionym: *Epidendrum vespa* Vellozo. 1827. *Florae Fluminensis* 9: t. 27.

DERIVATION OF NAME
Latin *vespa*, "wasp," referring to the plant's pollinator.

DESCRIPTION
An epiphyte with a long rhizome and glabrous, white roots. Pseudobulbs cylindric or spindle-shaped, compressed laterally, apex narrowed, initially covered by imbricating deciduous sheaths. Leaves two or three, oblong. Inflorescence emerging from the apex of the pseudobulbs. Floral bract triangular, acuminate. Flowers with short peduncle, sepals and petals greenish with purple asymmetric longitudinal lines; lip whitish with few purple veins in the apex; callus white. Dorsal sepal lanceolate, obtuse at apex. Lateral sepals elliptic, asymmetric, obtuse apex, a little shorter and wider than the dorsal sepal. Petals more or less falciform, acute at apex. Lip entire, cordiform (heart-shaped), hard and thick, glabrous; callus with two symmetrical rounded lamellas opposed to the column. Column gibbous, without auricles. Anther subcircular, retuse, yellow, two-lobed. Pollinarium with four yellowish pollinia. Stigmatic cavity more or less rectangular.

HABITAT AND DISTRIBUTION
Brazil (Atlantic Forest). Found in areas of high humidity, at 15–1000 meters in elevation.

FLOWERING TIME
In summer in Brazil.

COMMENT

For years, *Anacheilium vespa* had been thought to cover many types of vegetative and floral colors and patterns and a wide geographic spread. Part of the problem was that no original species *Epidendrum vespa* had been preserved for study by taxonomists. This made it hard to determine what parameters one would use to precisely define the species. All that changed when the true species was recently "rediscovered" by Marcos Campacci, who initially thought he had found a new species; however, on careful reading of the original description by Vellozo, Marcos realized he had found the authentic *Epidendrum vespa* Vellozo. He is quick to point out that the actual live plants were not lost at all; they have been growing there all along, the Brazilians just thinking they were an *Anacheilium vespa* of some sort. We use above the description he was going to use for his "new species."

With this species being now clearly defined, this leaves all the other listed "*vespas*" as some other species. In truth, there are probably several species still

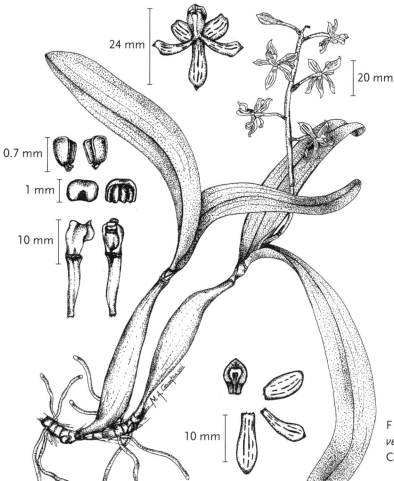

FIGURE 1-40. *Anacheilium vespa*. Drawing by Marcos Campacci.

mixed up in the leftovers, but we have put them in *Anacheilium crassilabium,* to be sorted out in the future. One might ask why we did not sort this out. To do so would require one to look at herbarium specimens throughout the world and to sample live populations. It will be a monumental task for someone in the future.

MEASUREMENTS
Pseudobulbs 8–12 cm long, 1.5–2 cm wide
Leaves 12–15 cm long, 2–3 cm wide
Inflorescence 15 cm long
Spathe 1.5 cm long
Sepals 15 mm long, 6 mm wide
Petals 15 mm long, 5 mm wide
Lip 8 mm long, 6 mm wide
Column 6 mm long, 4 mm wide

Anacheilium vinaceum

Anacheilium vinaceum (Christenson) Withner & Harding, *comb. nov.* Basionym: *Prosthechea vinacea* Christenson. 2003. *Richardiana* 3(3): 119. Type: Ecuador, Zamora-Chinchipe, beyond pass, Yangana to Valladolid, 2600 meters, 24 July 1985, *C. H. Dodson, Embree & Dalessandro 16033* (holotype: MO).

SYNONYMS
Encyclia pulcherrima (Klotzsch) Dodson & Bennett. 1989. *Icones Plantarum Tropicarum,* ser. 2 fasc. 1, t. 54.
Prosthechea pulchra Higgins, W. E. & Dodson. 2001. *Selbyana* 22 (2): 128, *nom. seminud.*

Not to be confused with
Epidendrum pulcherrimum Klotzsch. 1854. *Allgemeine Gartenzeitung* 22: 233; *Prosthechea pulcherrima* (Klotzsch) W. E. Higgins. 1997. *Phytologia* 82 (5): 381. This is a true *Epidendrum.*

DERIVATION OF NAME
Latin *vinaceus,* "wine-colored, purplish red," referring to the flower color.

DESCRIPTION
A terrestrial, growing in a tuft. Pseudobulbs slightly compressed. Leaves one or two. Inflorescence arising from spathe, flowers arranged in a pseudowhorl. Sepals and petals clear deep purple. Lip cerise-purple, callus platformlike with central

depression. Column greenish yellow, anther cap bright yellow. Sepals and petals are acuminate (pointed). Lip midlobe is much smaller than the lateral lobes.

HABITAT AND DISTRIBUTION
Endemic to Ecuador. Found at 2600 meters in montane wet forest.

FLOWERING TIME
July.

MEASUREMENTS
Pseudobulbs 6–12 cm long, 1 cm wide
Leaves 20 cm long, 1.5 cm wide
Inflorescence 10 cm long
Spathe 4 cm long
Sepals 8 mm long, 4 mm wide
Petals 8 mm long, 4 mm wide
Lip 6 mm long, 6 mm wide
Column 5 mm long, 3 mm wide

Anacheilium vita
FIGURE 1-41, PLATE 51

Anacheilium vita Hunt, Withner & Harding, *sp. nov.*
 Pseudobulbi 25 × 1.5 cm plus minusve compressi in rhizomate ad intervalla 8 cm separati. Folia tres, 28 × 2.5 cm. Inflorescentia ex vagina emergens. Pedi-

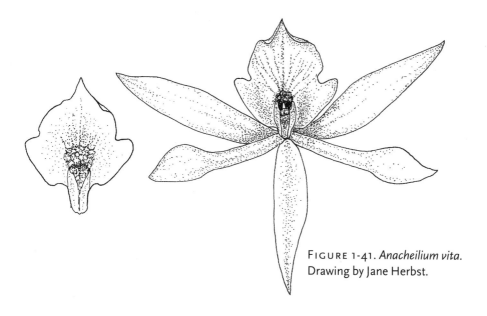

FIGURE 1-41. *Anacheilium vita*. Drawing by Jane Herbst.

cellus ovariumque 15 mm. Flores luteovirides, sepals petalisque lanceolatis 10–11 × 2–3 mm, labio convexo 7 × 5 mm luteo purpureo-guttato, callo bicarinato in medio sulcato apice purpureo, columna 5 mm longa unguiculata ad labium e basi plus minusve 1 mm connata. Species affines Anacheilium sceptro. Type: Ecuador, precise local unknown, *Hort. Ecuagenera s.n.* (holotype: US).

DERIVATION OF NAME
Latin *vita*, "life," in honor of Carl Withner and his generation of taxonomists.

DESCRIPTION
Pseudobulbs spindle-shaped, slightly compressed. Leaves two, linear-lanceolate. Inflorescence emerges from mature growths with a spathe. Flowers many (thirty-five to forty) green to yellow-green. Petals held fairly flat, not twisted or incurving. Lip white with dark maroon coloration extending from end of callus to near apex, leaving margins of lip free of maroon coloration. Lip three-lobed, though lateral lobes are small. Callus white, with two lobes and a central sulcus.

HABITAT AND DISTRIBUTION
Ecuador.

FLOWERING TIME
October.

MEASUREMENTS
Pseudobulbs 12–16 cm long, 1.6 cm wide
Leaves 21.5 cm long, 1.9 cm wide
Inflorescence 30–35 cm long
Sepals 11 mm long, 3 mm wide
Petals 10 mm long, 2 mm wide
Lip 7 mm long, 5 mm wide
Column 5 mm long, 2 mm wide

Anacheilium widgrenii
PLATE 50

Anacheilium widgrenii (Lindley) Pabst, Moutinho & A. V. Pinto. 1981. *Bradea* 3: 23. Basionym: *Epidendrum widgrenii* Lindley. 1853. *Folia Orchidacea Epidendrum* 39.

SYNONYMS

Hormidium widgrenii (Lindley) Brieger. 1961. *Publicacao Cientifica Universidade de São Paulo, Institut de Genetica*, 2: 69.

Encyclia widgrenii (Lindley) Pabst. 1976. *Bradea* 2 (14): 81.

Prosthechea widgrenii (Lindley) W. E. Higgins. 1997. *Phytologia* 82 (5): 381.

DESCRIPTION

Rhizome creeping. Pseudobulbs 1.5–2.5 cm apart, cylindrical, and slightly flattened. Leaves two. Flowers six to eight, non-resupinate, each 5 cm long, sepals lanceolate, tapering, petals more oval, both slightly reflexed with prominent midveins. Lip round, cochleate tapering at base of callus, ribbed along the axis. Flowers white to cream with dilution of rose. Lip has two calli on the section obscured by the column, lip streaked with purple near base.

HABITAT AND DISTRIBUTION

Brazil (Minas Gerais). Found at 500–1000 meters, with hot days and cool nights.

COMMENT

Pabst (1967), in *Orquídea*, lists *Anacheilium widgrenii* as a synonym of *Anacheilium papilio*. Reading the description of the two species it is hard to see much difference but looking at the pictures of both it is obvious that the stance on the petals is different, with the *A. widgrenii* being reflexed and *A. papilio* petals held more forward, and that the flowers of *A. widgrenii* are fully twice the size of those of *A. papilio*

MEASUREMENTS

Pseudobulbs 5–10 cm long, 0.7–2.5 cm wide
Leaves 8–25 cm long, 1.5–2.5 cm wide
Inflorescence 5–12 cm long
Spathe 3–6 cm long
Sepals 22–24 mm long, 5–6 mm wide
Petals 18 mm long, 8–9 mm wide
Lip 15–17 mm long, 12–13 mm wide
Column 6–8 mm long

Anacheilium undescribed species

FIGURE 1-42

DESCRIPTION

An epiphyte. Pseudobulbs clustered, stem cylindrical, thin. Leaves two. Peduncle 45 cm long. Flowers one to ten, non-resupinate. Sepals and petals yellowish green overlain with large mottled purple areas, with crested pleats running to apex. Lip white suffused with rose pink and purple and finely streaked with rose pink. Lip trilobed, lateral lobes transversely subquadrate, midlobe ovate, subacute, narrower than the lateral lobes. Callus a central low rectangular sulcate pad conforming to stigmatic area. Column greenish white, anther yellow, column short, stout, apex hooded, triangular, capping the anther, wings broad, with a subulate incurved apex. Ovary three-winged.

COMMENT

Figure 1-42 illustrates a plant from Peru that had been thought to be *Encyclia pulcherrima*, our *Anacheilium vinaceum*, until Dodson & Higgins (2001), in correcting the mistake in choosing *Epidendrum pulcherrimum* as a basionym species of *Prosthechea pulcherrima*, realized that it did not look like the type specimen. So now, the Peruvian "species" is unnamed, and to complicate matters, the Missouri Botanic Garden has misplaced the type specimens of both species. Until all this gets resolved, we leave the drawing of this Peruvian species here and its description. Figure 1-4 has drawings of both species' lips, and you can see that they are quite different. Both plants can be said to be similar to *Anacheilium hartwegii* but distinguished by brilliant color of the flowers, and the shapes of the lips.

MEASUREMENTS

Pseudobulbs 6–12 cm long, 1 cm wide
Leaves 20 cm long, 1.5 cm wide
Inflorescence 10 cm long
Spathe 4 cm long
Sepals 8 mm long, 4 mm wide
Petals 8 mm long, 4 mm wide
Lip 6 mm long, 6 mm wide
Column 5 mm long, 3 mm wide

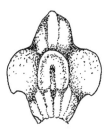
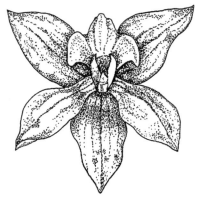

FIGURE 1-42. *Anacheilium* undescribed Peruvian plant. Drawing by Jane Herbst.

Anacheilium undescribed species

We have also obtained a picture, Plate 51, from Eric Christenson of a probably new and undescribed species he found in the greenhouses of Ecuagenera in Ecuador. Eric was sure, when he took the photo, it was new, but went back later to collect plant material and measurements and was unable to find the plant. You can see from the photograph that this is a very exciting and beautiful flower.

Anacheilium undescribed species

Type: PERU: Cajamarca, San Ignacio, District of San Jose de Lourdes, Crucero, 880 meters, *G. Calatayud 513* (holotype: CUZ).

Recently Gloria Calatayud has discovered a new, yet unnamed species from Peru, somewhat similar to *Anacheilium aloisii*. We will call it Calatayud #513. The two species differ by the flower color and the new species has sharply acutely pointed tepals. The new species is found in secondary forests. It is an epiphyte. Pseudobulbs are ellipsoid, 7–8 cm long. Leaves two, 20–23 cm long. Inflorescence is short with a peduncle 3–4 cm long. Flowers are yellow. Lip is unlobed, 6 mm long, 2.5 mm wide, has a disc with a pair of parallel callus keels. We have included a drawing of the lip in the collection of lip drawings.

Encyclia trautmannii
PLATE 52

Encyclia trautmannii Senghas. 2001. *Journal fur den Orchideenfreund.* 8 (1): 63. Type: N-Peru, ca. 1500 meters; leg. G. Trautmann, Institut für Systematische Botanik Heidelberg, sub no. Orch–913, 1997 (holotype: HEID).

DESCRIPTION
Pseudobulbs round. Leaves three. Lip long, smooth, and rimmed with a little callus at the base.

HABITAT AND DISTRIBUTION
This plant was reported to come from Peru, at 1500 meters

COMMENT

This is probably not a true species but a hybrid. Our first thought on seeing this plant was "Wow!" Our second thought was, "Is this a hybrid?" And our third thought was, "Why is this plant not widespread in cultivation?" Turns out that it is probably a case of a wrong tag in a pot, as the plant looks like *Epc.* Francis Dyer, which is a hybrid of *Cattleya bowringiana* and *Anacheilium fragrans.* We include the picture as our one example of a hybrid, and to keep the record clear.

CHAPTER 2

The Genus *Coilostylis*

In this chapter, we propose to resurrect the genus *Coilostylis* as a collection of species that will fill the *Epidendrum* subsection *Aulizeum* in Lindley's key. Flowers of the genus *Coilostylis* are uniformly similar, with their distinctive lateral lobes and narrow, long midlobe, and the flowers are born terminally, in small umbels or short racemes. They have pseudobulbs, though sometimes not obvious, which are very tough and thickened, bearing one or two leaves. We feel that these represent distinct and well-defined criteria for a genus.

Coilostylis (hollow style) has a complicated history dating back to 1836 when Rafinesque (*Flora Telluriana* 4: 37) first proposed the genus. In the same article, Rafinesque also published many other new generic names for orchids, and a variety of other taxa in different plant families. *Epidendrum emarginatum* was the type species for the genus *Coilostylis*; that name was later found a synonym of *Epidendrum ciliare*, a species that was named originally by Linnaeus in 1759. This old genus name might have remained in obscurity if Higgins, in defining the genus *Oestlundia,* had not discovered the name and thinking it a previous synonym of one of his species of *Oestlundia*, brought it to modern taxonomists' attention. It is the correct genus name by the rules of nomenclature if *Epidendrum ciliare* is separated from *Epidendrum* in the strict sense.

We had originally wanted to revitalize the name *Aulizeum* from Lindley's sections of *Epidendrum*. *Aulizeum* also has a complicated history and a variety of spellings. Salisbury had mentioned it as a genus in the Epidendrum group in 1812. It was also made a section of *Epidendrum* by Lindley in *Hooker's Journal of Botany* in 1841, and used as the section *Aulizodium* by Reichenbach f. in 1855. Lindley used it as a subgenus in his key of 1853. Bentham established the genus Aulizeum in 1881, and Stein (1892) used the name *Aulizeum ciliare* in his publications. During all of these changes, the epithets referred to *Epidendrum ciliare* Linnaeus as the type species.

We have seen the generic name *Auliza* in use in some technical orchid writings, but it has never been in general usage. The type species for the group is

Aulizeum ciliare, a binomial first published by Salisbury in 1812 (*Transactions of the Horticultural Society of London* 1: 294). In turn, the species was already known in pre-Linnaean polynomial times as *Helleborine graminea, foliis rigidis, carinatis* when it was listed in Plumier's 1703 *Plantarum Americanarum*. In 1758 when Burmann published Plumier's drawings, the polynomial name was printed as *Epidendrum foliis subradicalibus, oblongis, aveniis florum labio trifido, ciliato, intermedio lineari*. Since *Aulizeum* never was presented correctly in the literature nor did it enter general use, it cannot be the first valid generic name for the group of species that include *Epidendrum ciliare*.

If the subdivision of *Epidendrum* is to be carried through thoroughly, those orchid species that exist in the literature should be dealt with on some uniform basis. If one of Lindley's subgenera is raised to generic status, then the others with the same key characteristics should likely be dealt with in the same light. We are then faced with deciding which of the key characteristics are most critical and which should be given priority in this classification. We do not know why early investigators ignored this subgenus of Lindley's writings, or why the change to *Coilostylis* by Rafinesque was also made and ignored even more. Of course, they did not have a set of international rules to follow. Maybe they were just not willing to change a label, a factor that slows us down even today. Fortunately or unfortunately, *Coilostylis* has validity for this group by our present rules.

The plants of this genus have been known as the "eyelash epidendrums" because of the flowers of the type species, but there are only two species in the genus with fringed lateral lobes. Some are slightly fimbriated, but others are not fringed in any way. Another characteristic of the flowers is the tunnel-shaped column apex formed by the hooded clinandrium and throat of the internal spur. Fusion of the lip base with the column effectively internalizes and hides the style and stigma of the flower, as a longitudinal section through the column of this group of orchids will demonstrate. In this species, there is usually a pair of ridges or knoblike protrusions on the disc of lip midlobe that will guide an insect proboscis to the nectary. Other species of *Coilostylis* have the same lobed lip pattern with the midlobe of the lip distinctively long, linear, and pointed. As Lindley's key demonstrates, the lip is adnate to the column, a sheath or spathe subtends the flower stalk, and the plants are pseudobulbous. There was an initial temptation to make the genus large by including many *Epidendrum* species with white flowers (*Epidendrum longiflorum*, for example) with a long thin midlobe, but we will limit this group to encompass the parameters Lindley set, namely, plants with a pseudobulb.

These plants are easy to grow in any mixed collection, requiring almost no care. They tolerate long dry periods, do not appear to be sensitive to extremes in temperature (using a mixed culture orchid greenhouse temperature range as the norm), and reward the grower with blooms on a regular basis. They are best grown in a media that provides excellent drainage; a basket or mounted is best. They are sensitive to root disturbance, so once established it is best to leave them undisturbed.

Species of *Coilostylis*

Coilostylis ciliaris, 139
Coilostylis clavata, 141
Coilostylis cuspidata, 142
Coilostylis falcata, 143
Coilostylis lacertina, 144
Coilostylis oerstedii, 145
Coilostylis parkinsoniana, 145
Coilostylis vivipara, 146

Key to Species of *Coilostylis*

1a. Foliage flexible, oval or elliptic in shape . . . . go to 2
1b. Foliage markedly thickened, stiff and channeled, distinctly falcate (scythe-shaped). . . . . go to 7
2a. Leaf one, blooms on immature growth . . . . go to 3
2b. Leaves two or three, blooms on mature growth. . . . . go to 4
3a. Lateral lobes laterally toothed, sickle-shaped, midlobe with isthmus and spade-like, pointed end . . . . *Coilostylis clavata*
3b. Lateral lobes of the lip kidney-shaped, entire or most edged with rounded teeth, threadlike midlobe . . . . *Coilostylis oerstedii*
4a. Lip lateral lobes markedly fimbriate (toothed or fringed) . . . . go to 5
4b. Lip lateral lobes not fimbriate . . . . go to 6
5a. Leaves two, petals equal and curved, flowers greenish white . . . . . *Coilostylis ciliaris*
5b. Leaves three, petals stiff and linear, flowers yellow . . . . *Coilostylis cuspidata*
6a. Lip lateral lobes widest below the middle . . . . *Coilostylis vivipara*
6b. Lip lateral lobes widest basally. . . . . *Coilostylis lacertina*
7a. Pseudobulb obvious, plant crawling . . . . *Coilostylis falcata*
7b. Pseudobulb not obviously differentiated from rhizome, plant pendent . . . . *Coilostylis parkinsoniana*

Coilostylis ciliaris

PLATE 53

Coilostylis ciliaris (*Linnaeus*) Withner & Harding, *comb. nov.* Basionym: *Epidendrum ciliare* L. 1759. *Syst. Nat.* ed. 10, 1246.

SYNONYMS

Auliza ciliaris (Linnaeus) Salisbury. 1812. *Trans. Hort. Soc.* 1: 294, *nom. inval.*
Coilostylis emarginata Rafinesque. 1838. *Fl. Tellur.* 4: 37.
Epidendrum viscidum Lindley. 1840. *Edward's Botanical Register* 26, misc. 81.
Epidendrum cuspidatum var. *brachysepalum* Reichenbach f. 1847. *Linnaea* 19: 372.
Epidendrum ciliare var. *viscidium* Lindley. 1853. *Fol. Orch. Epid.* 30.
Epidendrum luteum Hort. 1858. *Planch. Hort. Don.* 165.
Epidendrum ciliare var. *minor* Hort. *ex* Stein. 1892. *Orchideenbuch* 226.
Phaedrosanthus ciliaris (Linnaeus) Kuntze. 1904. *Lexikon Generum Phanerogamarum* 429.
Epidendrum ciliare var. *squamatum* Schnee. 1953. *Rev. Fac. Afr.* 1: 206.

DERIVATION OF NAME

Latin *ciliare,* "ciliated," referring to the eyelash appearance of the lateral lobes.

DESCRIPTION

An epiphyte. Pseudobulbs light green, arising from a stout creeping rhizome, covered by a sheath when young. Leaf one, rarely two. Flowers four to six, white. Lip three-lobed, marked deeply with distinct longitudinal parallel furrows 5–8 mm long, lateral lobe half heart-shaped, sickle-shaped, extending forward, smooth on inner margin, deeply and irregularly sharp-fringed on outer margin. Midlobe needlelike. Disk with two erect flaps like keels extending from base of claw to the base of the midlobe, then with linear keels extending the length of the midlobe.

HABITAT AND DISTRIBUTION

Brazil, Colombia, Ecuador, México, Panama, Peru, West Indies, and Venezuela. Found at 550 to 2000 meters in elevation, commonly on rocks, but also epiphytic on trees.

FLOWERING TIME

April to June.

MEASUREMENTS

Pseudobulb 20–60 cm long
Leaves 5–23 cm long, 2–4.5 cm wide
Inflorescence 15–25 cm long
Sepals 40–90 mm long, 4–7 mm wide
Petals 40–80 mm long, 4–5 mm wide
Lip 17–40 mm long, 3–4 mm wide
Midlobes 25–60 mm long, 1 mm wide
Column 17–23 mm long

Coilostylis clavatum

PLATE 54

Coilostylis clavatum (Rafinesque) Withner and Harding, *comb. nov.* Basionym: *Didothion clavatum* Rafinesque. 1838. *Fl. Tellur.* 4: 39.

SYNONYMS
Epidendrum clavatum Lindley. 1836. *Bot Reg.* 22: t. 1870. non König 1791.
Epidendrum purpurascens Focke. 1851. *Tijdschr. Natuurk. Wetens. Nederl.* 4: 64.
Epidendrum glumibracteatum Reichenbach f. 1863. *Hamburger Garten- und Blumenzeitung* 19: 11.
Epidendrum clavatum var. *purpurascens* (Focke) Cogniaux. 1898. *Flora Brasiliensis (Martius)* 3 (5): 74.
Epidendrum psilanthemum Loefgren. 1917. *Arch. Jard. Bot. Rio de Janeiro* 2: 57.
Auliza clavatum (Lindley) Brieger. 1977. Schlechter's *Die Orchideen*, ed. 3, p. 545.

DERIVATION OF NAME
Latin *clavatus*, "club-shaped."

DESCRIPTION
An epiphyte. Pseudobulbs dark brown to olive green, often hard shiny, medium compressed but not two-edged. Sheath brown, very ragged. Leaf one. Inflorescence from new growth, peduncle compressed, concealed at base in large olive-brown bracts. Sepals pale green, petals somewhat paler green, lip white with two small white calli at base, adnate to apex of column. Column pale green at base grading to white at apex, anther white pollinia yellow.

HABITAT AND DISTRIBUTION
Brazil, Colombia, Costa Rica, French Guiana, Guyana, Surinam, and Venezuela. Found at 500 meters in elevation.

COMMENT
The original name, *Epidendrum clavatum*, was first given to an Old World plant, by J. König in 1791 (*Observationes Botanicae* 6: 50). That species is now considered assigned to genus *Dipodium*. Lindley used the name again, but once the name refers to a plant (species type), the name forever refers to that plant, so the later Lindley name is an illegitimate homonym. Rafinesque effectively validated the species name by publishing *Didothion clavatum* Rafinesque, which is treated as a *nom. nov.* and becomes the basionym for the species in *Coilostylis*. When speak-

ing of the species as an *Epidendrum* in the strict sense, the correct name is *Epidendrum purpurascens*.

MEASUREMENTS
Pseudobulb 5–8 cm long, 1.5–2.5 cm wide
Leaves 25 cm long, 3.5 cm wide
Sepals dorsal 20 mm long, 4.5 wide
Sepals lateral 19 mm long, 4.5 wide
Petals 19 mm long, 3 mm wide
Lip 10 mm long, 5 mm wide
Lateral lobes 10 mm across

Coilostylis cuspidata

Coilostylis cuspidata (Loddiges) Withner & Harding, *comb. nov.* Basionym: *Epidendrum cuspidatum* Loddiges. 1816. *Botanical Cabinet* 1: t. 10.

SYNONYM
Epidendrum ciliare var. *cuspidatum* (Reichenbach f.) Lindley. 1853. *Folia Orchidacea Epidendrum* 30.

DERIVATION OF NAME
Latin *cuspidatus*, "pointed," referring to the lip.

COMMON NAME
Yellow fringed *Epidendrum*.

DESCRIPTION
Flowers very short (we assume the describers meant that the inflorescence is short and that the flowers do not extend much above the foliage), large, pale straw in color. Lip has a deep yellow base.

HABITAT AND DISTRIBUTION
México and the West Indies.

COMMENT
This species can be confused with *Coilostylis ciliaris*, which has three leaves and petals that are equal and much curved. In *C. cuspidata*, the three sepals are equal and remarkably stiff and pointed, and the broad petals spread in the form of wings producing a distinctive appearance. The middle segment of the lip of *Coilostylis cuspidata* is quite linear, in *C. ciliaris* it is subulate (awl-shaped) and much longer.

The whole plant of *C. cuspidata* has a much larger growth form. The flower in *C. cuspidata* is yellow, in *C. ciliaris*, greenish.

Coilostylis cuspidata is often considered a synonym of *C. ciliaris* and may indeed be a subspecies of *C. ciliaris*.

Coilostylis falcata
PLATE 55

Coilostylis falcata (Lindley) Withner and Harding, *comb. nov.* Basionym: *Epidendrum falcatum* Lindley. 1840. *Annals of Natural History* 4: 382.

SYNONYMS
Epidendrum lactiflorum A. Richard & Galeotti. 1845. *Ann. Sc. Nat., sér.* 3, 3: 22.
Brassavola suaveolens Galeotti ex Hemsley. 1879. *Gard. Chron.* I, p.235.
Aulizeum falcatum Lindley ex Stein. 1892. *Orchideenbuch*, p.230.
Epidendrum parkinsonianum var. *falcatum* (Lindley) Ames, F. T. Hubbard & C. Schweinfurth. 1935. *Botanical Museum Leaflets* 3: 74.

DERIVATION OF NAME
Latin *falcatus*, "sickle-shaped," referring to the leaves.

DESCRIPTION
A lithophyte. Crawling plant with fleshy branching short stems. Pseudobulbs are easy to discern from the rhizome. Leaves long and channeled, sickle-shaped. Flowers four or more per growth emerging on a terminal, short, umbelliform raceme on a mature pseudobulb. Flowers are large, pale yellowish, and spring from within long pale yellowish green membranous sheaths. Flowers have a soapy smell at night.

HABITAT AND DISTRIBUTION
México. Found on limestone rocks and cliffs in pine and pine-oak forests, and xerophytic shrub and thorn forests at 1000–2100 meters in elevation.

FLOWERING TIME
Spring and summer.

COMMENT
The vegetative habit of this species is very different from its sister species *Coilostylis parkinsonianum*. The rhizome and pseudobulbs of this species are coarse, thick, and gnarly. It grows by creeping along a surface, whereas *C. parkin-*

sonianum rhizome and pseudobulbs are more delicate and orderly, and the plant dangles down from its base.

Coilostylis lacertina

Coilostylis lacertina (Lindley) Withner & Harding, *comb. nov.* Basionym: *Epidendrum lacertinum* Lindley. 1841. *Edward's Botanical Register* 27, misc. 53.

SYNONYMS
Epidendrum indusiatum Klotzsch. 1854. *Allgemeine Gartenzeitung* 22: 177.
    Type: Guatemala, *Warscewicz s.n.*
Auliza lacertinum (Lindley) Brieger. 1977. Schlechter's *Die Orchideen*, ed. 3, p. 547.

DERIVATION OF NAME
Latin *lacertina*, "little lizard," referring to shape of the lip.

DESCRIPTION
Leaves lanceolate, pointed. Flowers hang down on long stalked ovaries from one side of a congested raceme. Sepals lanceolate, pointed. Petals linear-lanceolate. Flowers bright green (or yellowish green and white) with the exception of the column which is yellow, lip slightly tinged with purple. Lip trilobed, lateral lobes lacinate, midlobe long pointed. Anthers hooded.

HABITAT AND DISTRIBUTION
Guatemala and México. Found at 2700 meters in elevation.

MEASUREMENTS
Plant up 240 cm long
Leaves 20 cm long, 3.5 cm wide
Inflorescence 12 cm long
Ovary 7 cm long
Sepals 22–45 mm long, 2.5–5 mm wide
Petals 20–40 mm long, 3 mm wide
Lip 17–25 mm long
Midlobe 13–20 mm long, 2 mm wide
Column 12–15 mm long

Coilostylis oerstedii
PLATE 56

Coilostylis oerstedii (Reichenbach f.) Withner & Harding, *comb. nov.* Basionym: *Epidendrum oerstedii* Reichenbach f. 1852. *Botanische Zeitung (Berlin)* 10: 937.

SYNONYMS
Epidendrum costaricense Reichenbach f. 1852. *Botanische Zeitung (Berlin)* 10: 937.
Epidendrum umlaufii Zahlbruckner. 1893. *Wien Illustr. Gartenzeit.* 18: 209. t. 2.
Epidendrum ciliare var. *oerstedii* (Reichenbach f.) L. O. Williams. 1946. *Annals of the Missouri Botanical Garden* 33: 327.

DERIVATION OF NAME
Honors the eminent Danish botanist and ecologist Anders Sandøe Øersted (1816–1872).

HABITAT AND DISTRIBUTION
Costa Rica and Panama. Found at 1150 meters in elevation.

COMMENT
The species is similar to *Coilostylis ciliaris* except the lateral lobes of the lip are kidney-shaped, entire or at most edged with rounded teeth, and the midlobe of the lip is threadlike to lancelike. It differs from *C. ciliaris* in that the lateral lobes are entire, and from *C. parkinsoniana* by having an erect, not pendent, habit.

Coilostylis parkinsoniana
PLATE 57

Coilostylis parkinsoniana (W. J. Hooker) Withner & Harding, *comb. nov.* Basionym: *Epidendrum parkinsonianum* W. J. Hooker. 1840. *Bot. Mag.* 67: t. 3778.

SYNONYMS
Epidendrum aloifolium Bateman. 1840. *Orch. Mex. Guat.* t. 25, non Linnaeus.
Epidendrum falcatum var. *zeledoniae* Schlechter. 1923. *Repertorium Specierum Novarum Regni Vegetabilis, Beihefte* 19: 37.
Epidendrum parkinsonianum var. *falcatum* (Lindley) Ames, F. T. Hubbard & C. Schweinfurth. 1935. *Botanical Museum Leaflets* 3: 74.

DERIVATION OF NAME
Honors the English botanist and herbalist John Parkinson.

DESCRIPTION
A large pendent branching epiphyte with a thick, short, decurved rhizome. Pseudobulbs present but not obvious unless you look for them, as they resemble the rhizome. Leaves one or two, slender, fleshy, leathery, and pendulous. Flowers terminal, two or three, very showy. Sepals and petals white, pale yellow or yellow green, lip white with pale orange mark in throat. Lip deeply three-lobed, midlobes obtriangular, outer margins dentate, midlobe linear. Disc with two flaplike keels.

HABITAT AND DISTRIBUTION
México, Guatemala, Honduras, Costa Rica and Panama. Found at 1700–1900 meters in elevation in humid forests, often along stream banks.

FLOWERING TIME
March and April.

COMMENT
Bateman used the name *Epidendrum aloifolium* for this species, but Linnaeus had already used that name for an Asian species now known as *Cymbidium aloifolium*.

MEASUREMENTS
Pseudobulb 6–10 cm long
Leaves 20–50 cm long, 1–3.5 cm wide
Inflorescence 1.5 cm long
Ovary 10–14 cm long
Sepals 56–85 mm long, 11–20 mm wide
Petals 55–80 mm long, 6–10 mm wide
Lip 55–80 mm long, 25–40 mm wide
Column 3 mm long

Coilostylis vivipara
PLATE 58

Coilostylis vivipara (Lindley) Withner and Harding, *comb. nov.* Basionym: *Epidendrum viviparum* Lindley. 1841. *Edward's Botanical Register*, misc. 10.

DERIVATION OF NAME
Latin *viviparum*, "bearing live young," referring to the plant's ability to generate young plants at the nodes.

DESCRIPTION

An epiphyte. Stems erect, becoming recumbent after flowering, elongated tapering to point at base, thickened toward apex. Stem surrounded by subfoliate sheaths. Leaves two or three. Flowers white, arranged in two vertical rows. Sepals and petals with inferior margin rolled backwards and downwards. Lip three-lobed, lateral lobes flat and thin, widest below the middle, somewhat heart-shaped at base, apical lobe long, unguiculate (narrowed into a claw or petiolelike base), flat and narrow, pointed. Callus formed of four fleshy lamellae, yellow, column white, apex having fine hairs, anther white, and pollinia two pair yellow. No fragrance.

HABITAT AND DISTRIBUTION

Bolivia, Brazil, Colombia, Ecuador, the Guianas, and Peru.

FLOWERING TIME

January to March.

COMMENT

Stem produces young plants at every node.

MEASUREMENTS

Stem 30–40 cm long
Leaves 15 cm long, 3 cm wide
Inflorescence 35 cm long
Ovary 4.5 cm long
Sepals 30 mm long, 5–6 mm wide
Petals 30 mm long, 5 mm wide
Lip 18 mm long, 15 mm wide
Column 20 mm long

CHAPTER 3

The Genus *Encyclia*

In this chapter Carl will continue briefly his discussion of the genus *Encyclia*, focusing on new opinions since the publication of his previous volumes including this genus (1988–2000).

The *Epidendrum* subgenus *Encyclia* in the Lindley key, the second subgenus to note, is today recognized as *Encyclia*, a genus proposed by Sir William Hooker in 1828 (before the key was written by Lindley) when Hooker published *Encyclia viridiflora* in Curtis's *Botanical Magazine*, t. 2831. Lindley accepted Hooker's genus and species, and therefore used it in his key, but as a subgenus. *Encyclia* has been used off and on in its generic status since those early days, and it has been so considered in Withner's various writing. Considering the species covered in volumes IV, V, and VI of *The Cattleyas and Their Relatives*, the genus now comprises, by Carl's count, 146 species, 27 from the Bahaman Islands and the Caribbean islands, 41 from Mesoamerica, 78 from South America, and one from Florida in the United States.

In this book, we note the probability of additional South American species that will have to be added to Withner's lists as further specimens from various countries are found. In particular, there seem to be at least a dozen new species forthcoming from Brazil, and other locations. There are, of course, various new opinions and species from Andean regions and other regions as well. *Encyclia dasilvae* is one example, published by V. P. Castro and M. A. Campacci in 2001 (*Icones Orchidacearum Brasilienses* 1: 54). The newest addition to the genus is *E. paraensis* from northeastern Brazil described by Castro and Cardoso in 2003 (*Richardiana* 3[2]: 69–73), a species close to *E. pachyantha* with identical colors but larger petal claws and a more compact rounder lip midlobe.

As additional species began to accumulate in his herbarium, Lindley subdivided *Encyclia* in four subgroups, and in 1853 (*Fol. Orch. Epi. Orch. Epi.*, p. 9), he published a key based on their distinguishing characteristics. The major features had to do with the lip morphology of the species, namely, if and how the lip was divided. The key is translated from Lindley's original Latin (correcting the translation from the Withner, vol. IV, p. 10):

1a. Labellum entire (unlobed) .... *Holochila*
1b. Labellum three-lobed at the apex, fleshy............................. *Sarcochila*
1c. Labellum three-lobed, midlobe lobe membranous, lateral lobes rotund, midlobe ± agreeing with, not larger .... *Sphaerochila*
2. Lateral lobes narrow, midlobe various, but always larger .... *Hymenochila*

The modern genera that were included in section *Epidendrum* subsection *Hymenochila* Lindley are those that were reviewed in previous volumes of *The Cattleyas and Their Relatives*, including *Artorima*, *Caularthron*, *Dimerandra*, *Dinema*, *Encyclia*, *Euchile*, and others. Subsection *Hymenochila* includes the species with membranous three-lobed lips, which we still consider *Encyclia* in the strict sense. They have a midlobe larger than the lateral lobes, and that distinguishes them from the *Sphaerochila*.

The type species for *Encyclia* is *E. viridiflora*, the first species of *Encyclia* to have been named. The type species for the subgenus *Hymenochila* is *Epidendrum tampense* Lindley (1847). This species was transferred to the genus *Encyclia* by Small in 1913. Withner (1996, 4: 11) discussed this and recognized *E. viridiflora* as the type species for the genus *Encyclia* and its subgenus *Encyclia*, and then validated *Encyclia* subgenus *Hymenochila* by recognizing *E. tampensis* as the type. The subgenus *Encyclia* has only a single species, *E. viridiflora*, and we may still use *Encyclia* as a genus name for specimens with the key characteristics of *E. tampensis*.

The key characteristics of *Encyclia* encompass vegetative as well as floral shapes, and once in mind, are easy to recognize, in particular the plant appearance. There is a pseudobulb, usually round, spherical, or somewhat pear-shaped, varying from oval to circular in cross section. The leaves are concentrated at the top of the pseudobulb, are usually stiff or leathery, and the number may vary from one or two, the most common arrangement, to three or even four leaves on a few species. The pseudobulbs may be smooth or longitudinally ridged, the latter if they have been subjected to much drying or if they are comparatively old or crowded. The pseudobulbs anatomically consist mostly of a single internode of the stem and have no leaf or bract scars, or enlargements, of their basic shape. They will produce a few bracts at the base, first green but then turning dry, papery, and peelable. Such pseudobulbs are called *heteroblastic*, as only one internode is really enlarged. The leathery leaves will be at the apex without apparent fleshy internodes between them. The leaves will be arched, or sometimes stiffly upright on top of the pseudobulb.

Other pseudobulbous *Epidendrum* may also have bracts and leaves, but they will be spaced along the stems at intervals, or they may have a series of enlargements, one above the other. Sometimes the pseudobulbs are fusiform (spindle-shaped), instead of being round or pyriform, and occasionally they are just stringy and fibrously tough, enlarged stems with leaves near their tops. These types of pseudobulbs are called *homoblastic* and are like those of *Broughtonia, Diacrium, Laeliopsis,* or *Schomburgkia*.

The flower stalk of *Encyclia* species arises from the very apex of the pseudobulb, and there is no particularly noticeable sheath (spathe) at its base. The stalk is either an unbranched raceme, or has a few to several branches (a panicle), and the age, culture, and condition of the plant can affect the kind of flower stalk, which may develop. Some small or young plants only produce racemes, but as the plants become larger, some may produce panicles. Plants of one species, *E. bracteata*, may produce only single flowers and for this reason alone may be considered by some as a distinct group.

Encyclia flowers have a lip that more or less encircles the column (the meaning of the Latin *encyclia*). The lip is three-lobed with elongated narrow lateral lobes, and it has a membranous midlobe that is usually larger than either of the lateral lobes. The anther protrudes at the tip of the column, covered with its anther cap, and is not surrounded by any teeth or low ridges such as are present on the column of the flowers of other Lindley sections of *Epidendrum*. This region around the anther cap is referred to as the clinandrium or anther bed.

The column is more or less oval or circular in cross section and does not possess a pronounced ridge along its back. There are no teeth at its apex, perhaps a point at most, separated from lateral points when present by broad shallow sinuses. The column may have membranous edges, and the edges may extend down in the region of the stigma as auricles that grasp the lip or callus ridges, augmenting the process of pollination. There is a nectary situated at the base of the column with its opening located at the base of the lip where it fuses with the column. The fruits are rounded and not three-angled (Dressler 1961). The callus is usually a forcipate (tweezerslike) structure—two fleshy ridges over the first pair of lateral veins alongside the central vein of the lip. The two ridges usually join over the central vein, as the veins extend from the callus area onto the disc of the midlobe. The central vein of the lip usually runs to the apex of the lip and often, with the pair of lateral veins on either side, forms a strong rib that helps support the midlobe.

If all the above conditions are true, the plant will be an *Encyclia*. These details are also discussed in volume IV *The Cattleyas and Their Relatives,* and most species of *Encyclia* are covered in volumes IV and V.

Six species are either newly described or their species status has been revised from past publications: *Encyclia brachiata, E. davidhuntii, E. dickdentii, E. kundergraberi, E. linearloba, E. mapiriensis.*

That leaves us with three of Lindley's sections not so far mentioned in this discussion: *Holochila, Sarcochila,* and *Sphaerochila.*

Encyclia species all have three-lobed lips. *Holochila* are characterized by an entire (unlobed) lip and this key character would place them into other genera. In his *Folia Orchidacea* (1853, p. 4), Lindley lists twelve species of *Encyclia* as belonging to this *Holochila* subgroup: *E. aureum* (synonym *Broughtonia aurea*), *E. auritum, E. brassavolae, E. distantiflorum, E. flabellatum* (synonym *E. candollei*), *E. ligulatum, E. lividum, E. luteoroseum, E. naevosum* (synonym *Barkeria naevosa*), *E. tripunctatum, E. subaquilum,* and *E. vitellinum.* These species are now dispersed in various genera and Lindley's subsection represents a polyphylic group of no taxonomic utility.

Just to throw in more information, Pabst in 1953 (Contribição Para o Conhecimento das Orquiídeas de Santa Catarina e Sua Dispersão Georgraphática–1, *Anais Botânicos do Herbário "Barbosa Rodrigues"* 5: 42) listed *Holochila* as a group under section *Aulizeum* of genus *Epidendrum* Linnaeus, and then lists *Epidendrum almasii, E. faustum, E. fragrans, E. papilio,* and *E. vespa* as species included. How this group of species is derived from Lindley's classification is not clear, or even if it is based on his classification, as we do not have the whole publication. We mention this here as just an example of alternate classification schemes that use the same sectional names. There are many examples in the literature.

Lindley lists subsection Sarcochila with four species: *Epidendrum glaucum, E. limbatum, E. ochraceum,* and *E. triste.* Of these, Lindley describes only *E. glaucum* as having the lip fleshy and three-lobed at the apex (*apice carnosi trilobi*). Thus, it must serve as the type species, and subsection *Sarcochila* becomes a straightforward synonym of *Prosthechea* in the strict sense.

Epidendrum subsection *Sphaerochila*, characterized by having three-lobed membranous lips, rounded lateral lobes, and the rounded midlobe not larger than the lateral lobes, is allotted four species by Lindley: *E. hastatum, E. prismatocarpum, E. pterocarpum,* and *E. squalidum. Epidendrum squalidum* cannot be the type because the midlobe is very fleshy (*crassissimo*). *Epidendrum pterocarpum* has a midlobe that is much longer than the lateral lobes (*multo longiore*), but midlobe is still smaller than the lateral lobes; as this might cause some confusion, it is better to pick another species for the lectotype. Of the remaining two species, *E. hastatum* appears to be the most suitable choice for a lectotype by having a roughly circular lip (*sphaerochila* meaning "globose lip"), unlike the lip of *E. prismatocarpum.* Thus, subsection *Sphaerochila* becomes a synonym of *Pollardia.*

CHAPTER 4

The Genus *Hormidium*

The genus *Hormidium* (Lindley) Heynhold 1841 (*Nomenclator Botanicus Hortensis* 1: 880) has been an enigma, it would seem to us, from the beginning. It has varied from a monotypic genus with only the type species to a large group in the recent Schlechter's *Die Orchideen,* third edition, with nine subgenera and about thirty-two sections. It has been converted into a sort of portmanteau genus where species have been transferred that did not fit into other more clearly defined generic concepts. Various investigators have tried their hand to define the genus more clearly, but it is still a puzzle.

The name *hormidium* means "to be strung like beads on a necklace," referring to the small beadlike pseudobulbs "strung" along a stringlike rhizome (Mayr 1998). In Lindley's original description of the *Epidendrum* subgenus *Hormidium,* and in his key, the flowers were described as sessile (without spathes or stalks), and with the lips adnate in some degree to the column. A ridge along the back of the column is distinct, and therefore the column in cross section is more triangular than oval or circular. The type species for this subgenus of *Epidendrum* was *E. uniflorum*. In the type description Lindley said, "A Mexican plant of no beauty, with yellowish green flowers, imported by George Barker, Esq. of Birmingham."

Lindley also places *Epidendrum pygmaeum* and *E. caespitosum* within this genus, which we now consider all synonyms of a single species. Lindley also stated that the species might be "defined with apparent precision by the union of the anterior (lateral) sepals with the base of the labellum." This fusion, in turn, forms a small cup or mentum at the base of the column. With this latter statement, he is describing characteristics of another species he placed in the subgenus, *E. miserum*. The traits of these two species combined muddled the definition of *Hormidium*. The characters he used are, in practice today, not as definitive as such a group of species would demand for generic distinction from other genera, and has resulted in several attempts at reviving the name and confusion about its use. Many of the epithets of the proposed subgenera and sections for this genus have not been validly published by the requirements of international code, further muddying this generic concept.

One account of *Hormidium*, by Cogniaux (in *Flora Brasiliensis* [*Martius*] 3 [5]: 28), emphasizes the small size of the plants, the fusiform shape of the pseudobulb, and the apparent lack of beauty of the flowers. The flowers are borne within a spathe and are usually considered sessile on the pseudobulbs. They do require close examination and patience with a lens to see the details properly. The plants are reputedly distributed from Brazil to Cuba and México.

Pabst, Moutinho, and Pinto (1981), also attempting to redefine *Hormidium*, listed the following key characters for the genus: absence of the crystal bundles characteristic of *Anacheilium*; labellum adnate to about half the column length; column distinctly ridged and enlarged or swollen (gibbous) toward the tip; an undivided rostellum; a three-winged or three-sided seed capsule; spindle-shaped pseudobulbs; an inflorescence without a spathe; flowers not resupinate, and frequent autogamy (self-pollination). Brieger et al. (1977) added that the ridge running along the back of the column, and the edges of the column that ordinarily terminate on either side of the clinandrium or anther bed, may be extended into three teeth or other protuberances. The ridge or edge of the anther bed may be plain or variously sculpted with serrations or other decorations.

We feel that Pabst characterized the genus well using the definitions above, except we question why he said that *Hormidium pygmaeum* has no spathe, because it clearly does. Other than that, we will use his definition of the genus.

Species of *Hormidium*

Hormidium pseudopygmaeum, 154 *Hormidium racemiferum*, 157
Hormidium pygmaeum, 155 *Hormidium rhynchophorum*, 158

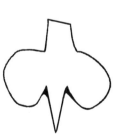

H. pseudopygmaeum   *H. pygmaeum*   *H. racemiferum*   *H. rhynchophorum*

FIGURE 4-1. *Hormidium* lips.

Key to Species of *Hormidium*

1a. Inflorescence surpassing leaves or subequal (about the same height as leaf tips) .... *Hormidium rhynchophorum*
1b. Inflorescence much shorter than leaves .... go to 2
2a. Floral bracts at least as wide as long, abruptly pointed, arranged on floral stem on alternating sides (two vertical sides) .... *Hormidium pseudopygmaeum*
2b. Floral bracts longer than wide, triangular-lanceolate, tapering gradually, spirally arranged on stem .... go to 3
3a. Flowers one to four, rachis (main axis of inflorescence) hidden, midtooth of column long, fingerlike .... *Hormidium pygmaeum*
3b. Flowers six to sixteen, rachis exposed, midtooth of column short, subequal (almost equal) to lateral teeth .... *Hormidium racemiferum*

Hormidium pseudopygmaeum

Hormidium pseudopygmaeum Finet. 1899. *Bulletin de l'Herbier Boissier* 7: 121, t. 3.

SYNONYMS

Encyclia pseudopygmaea (Finet) Dressler & G. E. Pollard. 1974. *Orquídea (México)* 3 (10): 310.

Prosthechea pseudopygmaea (Finet) W. E. Higgins. 1997. *Phytologia* 82 (5): 381.

DESCRIPTION

Pseudobulbs 6–10 cm apart on rhizome, narrowly spindle-shaped. Leaves two or three. Inflorescence short, with four to seven flowers. Sepals and petals basally cream shading to pale green. Lip cream with purple streak on midlobe. Lip adnate to column about half the length of column, three-lobed, lateral lobes oblong, clasping apex of column, midlobe triangular, pointed, callus of two low parallel oblong thickenings. Column midtooth narrowly oblong much longer than the lateral teeth. Capsule three-winged.

COMMENT

The species is similar to *Hormidium pygmaeum* but the plants are much bigger, the leaves are longer and narrower, and the flowers larger and more numerous. We find ourselves asking, "Why are these not just variations of the same species?" Traditionally these have been listed separately, and so we leave them separate, but you could think of them as the same species. Eric Christenson (pers. comm.) says "They are distinct but certainly horticulturally interchangeable."

HABITAT AND DISTRIBUTION
México and Central America to western Panama. Found at 1400–2700 meters in elevation, in wet or very wet cloud forest and mixed forest.

FLOWERING TIME
November to December.

CULTURE
We have no experience growing this plant. Culture should be the same as for *Hormidium pygmaeum*.

MEASUREMENTS
Pseudobulbs 4.5–10 cm long, 0.5–1 cm wide
Leaves 6–14.5 cm long, 1.2–1.6 cm wide
Inflorescence 1.7–3 cm long
Spathe 1–3.5 cm long
Sepals 8–13 mm long, 2–4 mm wide
Petals 7–9 mm long, 1.5–3 mm wide
Lip 6–8 mm long, midlobe 1–1.3 cm wide
Column 4–5.5 mm long

Hormidium pygmaeum
FIGURE 4-2, PLATE 59

Hormidium pygmaeum (W. J. Hooker) G. Bentham & J. D. Hooker. 1883. *Biol. Cent.–Am. Bot.* 3: 218. Basionym: *Epidendrum pygmaeum* W. J. Hooker. 1833. *Botanical Magazine* 60: t. 3233.

SYNONYMS
Coelogyne triptera Brongniart. 1834. *Dup. Voy. Coq. Phan.* 201, t. 42.
Epidendrum caespitosum Poeppig & Endlicher. 1838. *Nov. Gen. Ac. Sp.* 2: 1, t.101.
Epidendrum uniflorum Lindley. 1839. *Edward's Botanical Register* 25, misc. 16.
Epidendrum monanthum Steudel. 1840. *Nomencl. Bot.*, ed. 2, 1: 588.
Hormidium uniflorum (Lindley) Heynhold. 1841. *Nomencl. Bot. Hort.* 1: 880.
Hormidium tripterum (Brongniart) Cogniaux. 1898. *Flora Brasiliensis (Martius)* 3 (5): 50.
Microstylis humilis Cogniaux. 1906. *Flora Brasiliensis (Martius)* 3 (6): 550, t. 114.
Hormidium humile (Cogniaux) Schlechter. 1920. *Repertorium Specierum Novarum Regni Vegetabilis* 16: 331.

Encyclia pygmaea (W. J. Hooker) Dressler. 1961. *Brittonia* 13 (3): 265.
Encyclia triptera (Brongniart) Dressler & Pollard. 1971. *Phytologia* 21: 436.
Prosthechea pygmaea (W. J. Hooker) W. E. Higgins. 1997. *Phytologia* 82 (5): 381.

DERIVATION OF NAME
Latin *pygmaeus*, "pygmy, dwarf," referring to small size of the plant.

DESCRIPTION
Pseudobulbs 1–4 cm apart on long, scrambling rhizome, ovoid, slightly flattened. Leaves two or three. Inflorescence of one to three non-resupinate flowers. Sepals

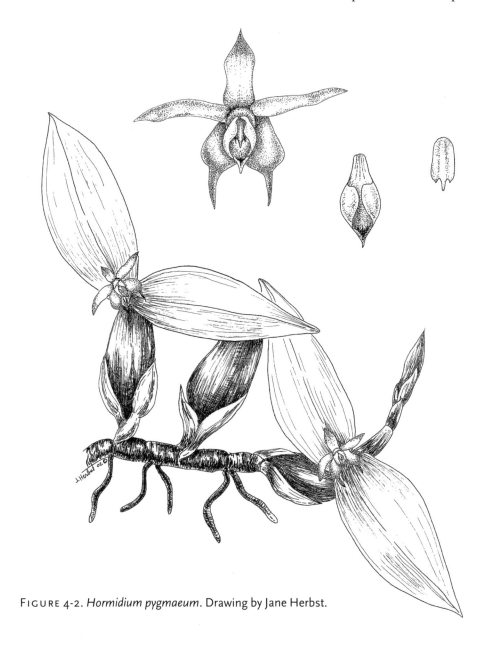

FIGURE 4-2. *Hormidium pygmaeum*. Drawing by Jane Herbst.

and petals cream or pale green, lip white with one to three purple spots or streaks in midlobe. Lip adnate with column for half its length, lip three-lobed, lateral lobes squarish clasping apex of column, midlobe triangular, pointed, callus of two low parallel oblong thickenings. Column midtooth lanceolate much longer than lateral teeth. Capsule three-winged.

HABITAT AND DISTRIBUTION

Widespread in the American tropics. Found at 200–2200 meters in elevation, in rather dry tropical forests and wet oak forests at higher elevations, often found on forest floor or on rocks.

FLOWERING TIME

November and December in México, October to April in Jamaica.

CULTURE

Culture for this plant is intermediate to warm temperatures, wet most of the year. We grow it mounted, as it grew out of a basket almost immediately.

MEASUREMENTS

Pseudobulbs 1.8–3 cm long, 5–9 cm wide
Leaves 1.5–14 cm long, 0.8–2 cm wide
Inflorescence 4–5 cm long
Spathe 3–6 cm long,
Sepals 5–11 mm long, 1.5–4 mm wide
Petals 4–9 mm long, 0.8– 2 mm wide
Lip 4–9 mm long, midlobe 0.8–1 mm wide
Lateral lobe 3.2–8 mm wide
Column 2–5 mm long

Hormidium racemiferum

Hormidium racemiferum (Dressler) Withner & Harding, *comb. nov.* Basionym: *Encyclia racemifera* Dressler. 1996. *Lindleyana* 11 (1): 37. Type: Costa Rica, Province of San José, vicinity of El General, 1470 meters, August 1936, *A. F. Skutch 2807* (holotype: AMES; isotype: MO).

SYNONYM

Prosthechea racemifera (Dressler) W. E. Higgins. 1997. *Phytologia* 82: 380.

DERIVATION OF NAME

Latin *racemi*, "bearing a raceme," referring to the inflorescence.

DESCRIPTION

Epiphyte or lithophyte. Pseudobulbs ellipsoid. Leaves two or three. Inflorescence a raceme of six to fifteen flowers. Sepals green or brownish green, petals same or white. Lip white with purple spot on midlobe. Lateral lobes ovate, parallel with lip axis, midlobe triangular acute. Column with midtooth longer than lateral teeth.

HABITAT AND DISTRIBUTION

Costa Rica. Found at 600–1470 meters in elevation.

FLOWERING TIME

August.

COMMENT

This species may be confused with *Hormidium pygmaeum* if one ignores the relatively large and many-flowered inflorescence.

MEASUREMENTS

Pseudobulbs 2.5–6 cm long, 0.4–0.9 cm wide
Leaves 2–8 cm long, 0.7–1.3 cm wide
Inflorescence 2–4.5 cm long
Spathe 0.7–1.7 cm long
Sepals 5–7 mm long, 1.5–2 mm wide
Petals 5–5.5 mm long, 0.7–0.9 mm wide
Lip 4–4.8 mm long, midlobe 1–1.8 mm wide
Column 3 mm long

Hormidium rhynchophorum

FIGURE 4-3, PLATE 60

Hormidium rhynchophorum (A. Richard & Galeotti) Withner & Harding, comb. nov. Basionym: *Epidendrum rhynchophorum* A. Richard & Galeotti. 1845. *Annales des Sciences Naturelles; Botanique*, sér. 3, 3: 20.

SYNONYMS

Encyclia rhynchophora (A. Richard & Galeotti) Dressler. 1961. *Brittonia* 13: 265.
Prosthechea rhynchophora (A. Richard & Galeotti) W. E. Higgins. 1997. *Phytologia* 82 (5): 381.

DERIVATION OF NAME

Latin *rynchos*, "beak," and *phora*, "bearer," referring to the lip.

DESCRIPTION

Pseudobulbs ovoid. Leaves one or two. Flowers three to six but usually four or five, resupinate. Sepals and petals green-yellow with some light brown spots. Lip three-lobed, deep green-yellow, with pale brown spots, the lateral lobes ending very abruptly as if cut straight across, clasping the column, middle lobe narrow long point.

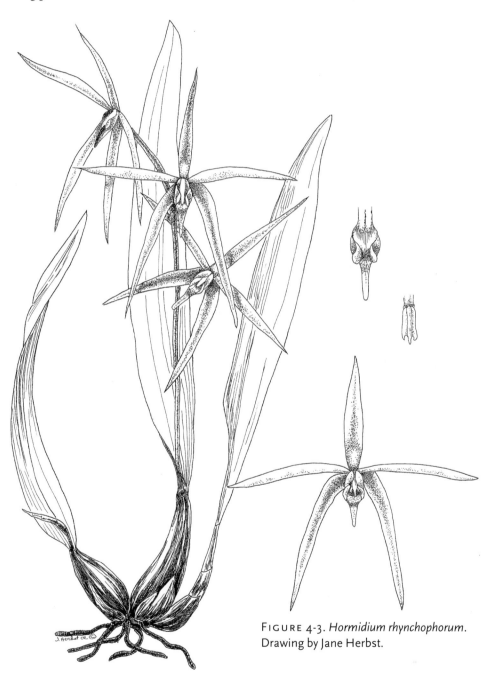

FIGURE 4-3. *Hormidium rhynchophorum*. Drawing by Jane Herbst.

HABITAT AND DISTRIBUTION
El Salvador, Honduras, México, and Nicaragua. Found at 2000 meters in elevation.

FLOWERING TIME
November and March to April.

CULTURE
Patricia grows this plant mounted, in intermediate conditions, where it gets year-round moisture though less in the winter. The plant was obtained from Weyman Bussey who recommended this culture. It seems to be growing well with these conditions.

MEASUREMENTS
Pseudobulbs 4 cm long, 2.8 cm wide
Leaves 18 cm long, 1 cm wide
Inflorescence 10 cm long
Sepals 15–27 mm long, 2 mm wide
Petals 14 mm long, 1.2 mm wide
Lip 7 mm long, 1 mm wide
Column 4 mm long

PLATE 1. *Anacheilium abbreviatum*. Photo by Leon Glicenstein.

PLATE 3. *Anacheilium alagoense*. Photo by Marcos Campacci.

PLATE 2. *Anacheilium aemulum*. Photo by Manfred Speckmaier.

PLATE 4. *Anacheilium allemanii*. Photo by Robert Anderson.

PLATE 5. *Anacheilium allemanii*. Photo by Marcos Campacci.

PLATE 6. *Anacheilium allemanoides*. Photo by Ron Parsons.

PLATE 7. *Anacheilium baculus*. Photo by David Hunt.

PLATE 8. *Anacheilium bennettii*. Photo by Carlos Hajek.

PLATE 9. *Anacheilium brachychilum*, another color form. Photo by Eric Christenson.

PLATE 10. *Anacheilium brachychilum*. Photo by Eric Christenson.

PLATE 11. *Anacheilium bulbosum*. Photo by Robert Anderson.

PLATE 12. *Anacheilium caetense*. Photo by Marcos Campacci.

PLATE 13. *Anacheilium calamarium* Photo by Manfred Speckmaier.

PLATE 14. *Anacheilium campos-portoi*. Photo by Marcos Campacci.

PLATE 15. *Anacheilium carrii*. Photo by Marcos Campacci.

PLATE 16. *Anacheilium chacaoense*. Photo by Manfred Speckmaier.

PLATE 17. *Anacheilium chimborazoense*. Photo by Manfred Speckmaier.

PLATE 18. *Anacheilium chondylobulbon*. Photo by Manfred Speckmaier.

PLATE 19. *Anacheilium cochleatum*. Photo by Patricia Harding.

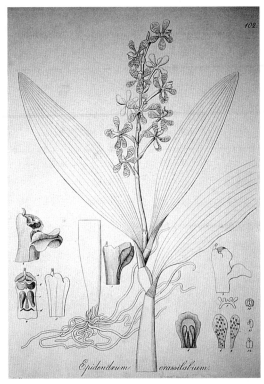

PLATE 20. *Anacheilium crassilabium*, drawing of type. Photo by Eric Christenson.

PLATE 21. *Anacheilium crassilabium*. Photo by Rudolph Jenny.

PLATE 22. *Anacheilium crassilabium*. Photo by Patricia Harding.

PLATE 24. *Anacheilium faustum*. Photo by Marcos Campacci.

PLATE 23. *Anacheilium faresianum*. Photo by Ron Parsons.

PLATE 25. *Anacheilium fragrans*. Photo by Ron Parsons.

PLATE 26. *Anacheilium fuscum*. Photo by Carlos Hajek.

PLATE 27. *Anacheilium garcianum*. Photo by Robert Anderson.

PLATE 28. *Anacheilium gilbertoi*. Photo by Ron Parsons.

PLATE 29. *Anacheilium glumaceum*. Photo by Manfred Speckmaier.

PLATE 30. *Anacheilium hajekii*. Photo by Carlos Hajek.

PLATE 31. *Anacheilium hartwegii*. Photo by Rudolf Jenny.

PLATE 32. *Anacheilium ionophlebium*? Photo by David Hunt.

PLATE 33. *Anacheilium janeirense*. Photo by Marcos Campacci.

PLATE 34. *Anacheilium kautskyi*. Photo by Roberto Anselmo Kautsky.

PLATE 35. *Anacheilium lindenii*. Photo by Manfred Speckmaier.

PLATE 36. *Anacheilium mejia*. Photo by Masa Tsubota.

PLATE 37. *Anacheilium papilio*. Photo by Mauro Peixoto.

PLATE 38. *Anacheilium radiatum*. Photo by Robert Anderson.

PLATE 39. *Anacheilium santanderense*. Photo by David Hunt.

PLATE 40. *Anacheilium sceptrum*. Photo by Manfred Speckmaier.

PLATE 41. *Anacheilium simum*. Photo by Ron Parsons.

PLATE 42. *Anacheilium spondiadum*. Photo by Leon Glicenstein.

PLATE 43. *Anacheilium tigrinum*. Photo by Ron Parsons.

PLATE 44. *Anacheilium trulla*. Photo by Manfred Speckmaier.

PLATE 45. *Anacheilium vagans*. Photo by Leon Glicenstein.

PLATE 46. *Anacheilium vasquezii*. Photo by Carlos Hajek.

PLATE 47. *Anacheilium venezuelanum*. Photo by Manfred Speckmaier.

PLATE 48. *Anacheilium vespa*, showing what was to be a new species *"A. lineatum"* but in reality is *A. vespa* and the one labeled *vespa* is *A. crassilabium*. Photo by Marcos Campacci.

PLATE 49. *Anacheilium vita*. Photo by Carl Withner.

PLATE 50. *Anacheilium widgrenii*. Photo by Manfred Speckmaier.

PLATE 51. *Anacheilium* sp. nov. Photo by Eric Christenson.

PLATE 52. *Anacheilium trautmannii*—?*Epc.* Francis Dyer. Photo by Karlheinz Senghas.

PLATE 53. *Coilostylis ciliaris*. Photo by Dale Borders.

PLATE 54. *Coilostylis clavata*. Photo by Dale Borders.

PLATE 55. *Coilostylis falcata*. Photo by Andy's Orchids.

PLATE 56. *Coilostylis oerstedii*. Photo by Carl Withner.

PLATE 57. *Coilostylis parkinsoniana*. Photo by Andy's Orchids.

PLATE 59. *Hormidium pygmaeum*. Photo by Greg Allikas.

PLATE 58. *Coilostylis vivipara*. Photo by Dwayne Lowder.

PLATE 60. *Hormidium rhynchophorum*. Photo by Rudolph Jenny.

PLATE 61. *Oestlundia cyanocolumna*. Photo by Rudolph Jenny.

PLATE 62. *Oestlundia luteorosea*. Photo by Weyman Bussey.

PLATE 63. *Oestlundia tenuissima*. Photo by Andy's Orchids.

PLATE 64. *Panarica brassavolae*. Photo by Dale Borders.

PLATE 65. *Panarica ionocentra*. Photo by Leon Glicenstein.

PLATE 66. *Panarica prismatocarpa*. Photo by Leon Glicenstein.

PLATE 67. *Panarica prismatocarpa*. Photo by Mauro Peixoto.

PLATE 68. *Pollardia campylostalix*. Photo by Ron Parsons.

PLATE 69. *Pollardia concolor*. Photo by Rudolph Jenny.

PLATE 70. *Pollardia ghiesbreghtiana*. Photo by Ron Parsons.

PLATE 71. *Pollardia hastata*. Photo by Manfred Speckmaier.

PLATE 72. *Pollardia linkiana*. Photo by Manfred Speckmaier.

PLATE 73. *Pollardia livida* Mexican variety. Photo by Robert Anderson.

PLATE 74. *Pollardia livida* Venezuelan variety. Photo by Manfred Speckmaier.

PLATE 75. *Pollardia michuacana*. Photo by Manfred Speckmaier.

PLATE 76. *Pollardia obpiribulbon*. Photo by Dale Borders.

PLATE 77. *Pollardia pringlei*. Photo by Andy's Orchids.

PLATE 78. *Pollardia pterocarpa*. Photo by Weyman Bussey.

PLATE 79. *Pollardia semiaptera*. Photo by Patricia Harding.

PLATE 80. *Pollardia tripunctata*. Photo by Manfred Speckmaier.

PLATE 81. *Pollardia varicosa*. Photo by Ron Parsons.

PLATE 82. *Pollardia venosa* another variety. Photo by Ron Parsons.

PLATE 83. *Pollardia venosa* one variety. Photo by Manfred Speckmaier.

PLATE 84. *Prosthechea bicamerata*. Photo by Andy's Orchids.

PLATE 85. *Prosthechea boothiana*. Photo by Greg Allikas.

PLATE 86. *Prosthechea brachiata*. Photo by Carl Withner.

PLATE 87. *Prosthechea glauca*, drawing. Photo by Rudolph Jenny.

PLATE 88. *Prosthechea glauca*. Photo by Ron Parsons.

PLATE 89. *Prosthechea grammatoglossa*. Photo by Marcos Campacci.

PLATE 90. *Prosthechea guttata*. Photo by Ron Parsons.

PLATE 92. *Prosthechea ochracea*. Photo by Manfred Speckmaier.

PLATE 91. *Prosthechea magnispatha*. Photo by Dwayne Lowder.

PLATE 93. *Prosthechea panthera*. Photo by Dan and Marla Nikirk.

PLATE 95. *Epidendrum bracteolatum*. Photo by Trudy Marsh.

PLATE 94. *Prosthechea vitellina*. Photo by Ron Parsons.

PLATE 96. *Epidendrum stamfordianum*. Photo by Trudy Marsh.

PLATE 97. *Epidendrum stamfordianum.* Photo by Manfred Speckmaier.

PLATE 98. *Epidendrum stamfordianum.* Photo by Manfred Speckmaier.

CHAPTER 5

The Genus *Oestlundia*

Best described as a genus of small plants with narrow leaves, *Oestlundia* was first defined by Dressler and Pollard in 1971 as a section of *Encyclia* rather than as a genus. This section has essentially been ignored since its publication, and few orchid growers appear to have noted it in the literature or found the plants available for cultivation. The taxon, nevertheless, fits well into the scheme originated by Lindley for the subsections of *Epidendrum* and has been raised to generic status.

This genus has now been named *Oestlundia* for Karl Erik Magnus Östlund (1857–1938), the collector of the type specimen of the type species. Higgins, in his work on *Epidendrum*, resolved the issue of this small group of plants. Traditionally *E. subulatifolium* has been included in the group, but now clearly can be excluded. The holomorphological cladogram that Higgins produced for this group of plants shows them to be within the subtribe Laeliinae, but they do not belong within *Encyclia* and are now considered a distinct genus.

The species of *Oestlundia* are found in Central America, primarily México, though one species at least may have an extended distribution that can include Peru.

The pseudobulbs of *Oestlundia* species are clustered, conical-ovoid. There are two or three grasslike leaves per pseudobulb. Inflorescences are simple or branched, with small flowers that are large for the size of plant. The lip blade has low keels that become warty distally. The plants have a lip that is adnate to the column with a smooth lip transition to the column; the column has a rounded midtooth much shorter than winglike lateral teeth which surpass the anther. The flowers characteristically have a short column and comparatively long winglike lateral lobes on the lip that extends toward the column tip. The columns do not produce auricles extending laterally that are more typical on the columns in the genus *Encyclia*. The seed capsule is ellipsoid.

Plants of this genus tend to make matlike growths with nearly spherical small pseudobulbs. When they are in flower on a slab of treefern or cork, the flowers can produce a nice show. Because of their small size and narrow leaves, the plants are dependent upon a reasonably good and constant supply of humidity during growth and flowering. When dormant, the plants may be kept on the drier side. They have the general habits of other twig epiphytes, particularly those in the genus *Encyclia*, and seem sensitive to change. They are good subjects for the experienced grower.

In combining and sorting species contained in this book, we note that the column teeth of *Prosthechea arminii, P. grammatoglossa,* and *P. vitellina* have the same characteristics as the column teeth of *Oestlundia,* though the remaining characteristics are dissimilar.

Species of *Oestlundia*

Oestlundia cyanocolumna, 202
Oestlundia distantiflora, 203
Oestlundia luteorosea, 204
Oestlundia tenuissima, 205

Key to Species of *Oestlundia*

1a. Petals less than or equal to 1 mm wide . . . . go to 2
1b. Petals wider than 1 mm . . . . go to 3
2a. Column yellow, orange-yellow, green-yellow, but most characteristically dark purple or blue-violet . . . . *Oestlundia cyanocolumna*
2b. Column cream white with faint purple veining and edging . *Oestlundia distantiflora*
3a. Lip white . . . . *Oestlundia luteorosea*
3b. Lip yellow. . . . . *Oestlundia tenuissima*

Oestlundia cyanocolumna
PLATE 61

Oestlundia cyanocolumna (Ames, F. T. Hubbard & C. Schweinfurth) W. E. Higgins. 2001. *Selbyana* 22: 4. Basionym: *Epidendrum cyanocolumnum* Ames, F. T. Hubbard & C. Schweinfurth. 1934. *Botanical Museum Leaflets* 3: 2. Type: México, State of Tamaulipas, near Jaumave, 13 June 1932, leg. H. W. v. Roszinsky *Eric Östlund 668* (holotype: AMES, no. 39432).

SYNONYMS
Encyclia cyanocolumna (Ames, F. T. Hubbard & C. Schweinfurth) Dressler. 1961. *Brittonia* 13 (3): 264.
Hormidium cyanocolumnum (Ames, F. T. Hubbard & C. Schweinfurth) Brieger. 1977. Schlechter's *Die Orchideen*, ed. 3, p. 574, as *Hormidium cyanocolumna.*

DERIVATION OF NAME
Latin *cyano,* "blue," and *columna,* "column," referring to the color of the column.

DESCRIPTION
Pseudobulbs clustered, conical-ovoid. Leaves two or three, grasslike. Inflorescence can be simple or branched. Flowers five to many. Sepals and petals olive green to yellow, lip white to white-yellow with violet spots or stripe centrally. Column yellow, orange-yellow, green-yellow, dark purple or blue-violet. Lip adnate to column for about a third of the length. Lip distal margin wavy. Callus two short keels at base, lip blade has five to seven low keels which become warty distally. Column midtooth rounded, much shorter than winglike lateral teeth that surpass the anther. Capsule ellipsoid.

HABITAT AND DISTRIBUTION
México. Found at 1500–2000 meters in elevation in oak forests.

FLOWERING TIME
April to June.

CULTURE
Culture is intermediate to cool temperatures, with semishade to moderate indirect light.

COMMENT
This is the type species for the genus *Oestlundia*.

MEASUREMENTS
Pseudobulbs 1–1.7 cm long, 0.5–0.8 cm wide
Leaves 4–10 cm long, 0.2–0.3 cm wide
Inflorescence 4–30 cm long
Sepals 7.5–9.5 mm long, 2–2.5 mm wide
Petals 7.5–9.5 mm long, 0.5–0.7 mm wide
Lip 10–15 mm long, 4.5–6 mm wide
Column 4.5–5 mm wide

Oestlundia distantiflora

Oestlundia distantiflora (A. Richard & H. Galeotti) W. E. Higgins. 2001. *Selbyana* 22: 4. Basionym: *Epidendrum distantiflorum* A. Richard & H. Galeotti. 1845. *Annales des Sciences Naturelles; Botanique,* sér. 3, 3: 19.

SYNONYM
Encyclia distantiflora (A. Richard & H. Galeotti) Dressler & G .E. Pollard. 1971. *Phytologia* 21 (7): 437.

DERIVATION OF NAME
Latin *distans,* "standing apart," and *flora,* "flowers," referring to the stance of the flowers on the inflorescence.

DESCRIPTION
Pseudobulbs clustered, conic-ovoid. Leaves one to three. Inflorescence branched, many flowered. Sepals and petals pale yellow or greenish yellow, lip white with green median stripe. Lip basally adnate to column, pointed, blade of lip narrow, lance-shaped, undivided. Callus two short keels at base of lip blade, blade with seven low keels, median keel reaches apex. Column cream white with faint purple veining and edging. Column midtooth blunt, shorter than winglike lateral teeth which surpass the anther.

HABITAT AND DISTRIBUTION
México and Belize. Found at 900–1000 meters in elevation in pine forest.

FLOWERING TIME
July to September.

MEASUREMENTS
Pseudobulbs 1–2.5 cm long, 0.8–1.5 cm wide
Leaves 9–19 cm long, 0.3–0.7 cm wide
Inflorescence 17–40 cm long
Sepals 16–18 mm long, 2 mm wide
Petals 16–17 mm long, 0.8–1 mm wide
Lip 15–16 mm long, 3.5–4 mm wide
Column 3–3.5 mm long

Oestlundia luteorosea
PLATE 62

Oestlundia luteorosea (A. Richard & H. Galeotti) W. E. Higgins. 2001. *Selbyana* 22: 4. Basionym: *Epidendrum luteoroseum* A. Richard & H. Galeotti. 1845. *Annales des Sciences Naturelles; Botanique,* sér. 3, 3: 19.

SYNONYMS
Epidendrum lineare Ruíz & Pavón. 1798. *Systema Vegetabilium Florae Peruvianae et Chilensis* 1: 249. nom. illeg., non Jacquin 1760.
Epidendrum seriatum Lindley. 1853. *Folia Orchidacea Epidendrum* 59.
Encyclia linearis Dressler. 1961. *Brittonia* 13 (3): 265. 1961. nom. illeg.
Encyclia luteorosea (A. Richard & H. Galeotti) Dressler & G. E. Pollard. 1971. *Phytologia* 21: 437.

DERIVATION OF NAME

Latin *luteo,* "yellow," and *rosea,* "of roses."

DESCRIPTION

Pseudobulbs loosely clustered on rhizome, ovoid to spindle-shaped. Leaves two to four. Inflorescence branched and many flowered. Sepals and petals green-yellow shading distally to brown or purplish brown, lip cream white centrally marked with dull violet. Sepals oblanceolate, either pointed or rounded, petals spatulate, rounded. Lip adnate to column for three-fifths the length of the column, egg-shaped attached at narrow end, the apex bluntly rounded with a notch. Callus two fleshy ridges passing into 3 or 5 very fleshy warty veins that run to apex of lip. Column purplish with teeth midtooth bluntly rounded, shorter than the winglike lateral teeth. Capsule ellipsoid.

HABITAT AND DISTRIBUTION

México, Central America, Peru, and Venezuela. Found at 600–2000 meters in elevation in dry oak or scrub oak forest, but sometimes in rather wet forests.

COMMENT

The epithet *lineare* was used by Jacquin in 1760 for *Epidendrum lineare* (*Selectarum Stirpium Americanarum Historia.* 221, t. 131, f. 1), making *E. lineare* Ruíz & Pavón an invalid name.

MEASUREMENTS

Pseudobulbs 2–7 cm long, 0.8–2.5 cm wide
Leaves 13–25 cm long, 0.5–1 cm wide
Inflorescence 8–45 cm long
Spathe 0.7–2.5 cm long
Sepals 7–10 mm long, 2–3 mm wide
Petals 7–11 mm long, 1.5–3 mm wide
Lip 8–9 mm long, 4–5 mm wide
Column 5 mm long

Oestlundia tenuissima

PLATE 63

Oestlundia tenuissima (Ames, F. T. Hubbard & C. Schweinfurth) W. E. Higgins. 2001. *Selbyana* 22: 4. Basionym: *Epidendrum tenuissimum* Ames, F. T. Hubbard & C. Schweinfurth. 1934. *Botanical Museum Leaflets* 3: 15. Type: México, State of Michoacán, 2 May 1933. *Erik M. Östlund 2246* (O. Nagel) (holotype: AMES, no. 39911).

SYNONYM
Encyclia tenuissima (Ames, F. T. Hubbard & C. Schweinfurth) Dressler. 1961. *Brittonia* 13 (3): 265.

DERIVATION OF NAME
Latin *tenuissima*, "the most slender," referring to the foliage.

DESCRIPTION
Pseudobulbs clustered, ovoid or spherical ovoid, ellipsoid. Leaves two or three. Inflorescence simple or few-branched. Flowers two to twelve. Sepals elliptic-oblong, petals oblanceolate, pointed or rounded. Flowers yellow (butter yellow) to orange-yellow. Lip adnate to column for three-fifths the length of the column. Lip blade unlobed, egg-shaped, callus of two short keels, lip blade covered by fleshy veins, median veins very warty, lateral veins toothed (merely crenulate/scalloped), three on each side. Column midtooth rounded, much shorter that the winglike lateral teeth. Rostellum forms a semiliquid "viscidium" which is removed with the pollinia as a unit (Dressler 1974). Capsule ellipsoid.

HABITAT AND DISTRIBUTION
México. Found at 1800–200 meters in elevation, in rather dry scrubby oak forest.

FLOWERING TIME
February to June.

COMMENT
This species is related to *Oestlundia luteorosea* but differs in that *O. luteorosea* has five rows of papillae, warty ridges, and is without a broad warty band in the center unlike *O. tenuissima*, which has a warty median vein.

MEASUREMENTS
Pseudobulbs 0.7–2.2 cm long, 0.3–1 cm wide
Leaves 3.5–10.7 cm long, 0.15–4 cm wide
Inflorescence 5–15 cm long,
Sepals 8.8–13 mm long, 1.5–3 mm wide
Petals 8–12 mm long, 1.4–2.3 mm wide
Lip 7–13 mm long, 4–7 mm wide
Column 5–6 mm long

CHAPTER 6

The Genus *Panarica*

Franco Pupulin recently reviewed this group in Costa Rica, described two new species, and characterized the salient features of the group. Pupulin described the complex as having pyriform pseudobulbs, two or rarely three leaves, showy flowers on erect inflorescences, flowers with three-lobed lips (except *Panarica brassavolae*) with the median lobe larger than the lateral lobes, the median lobe acute (pointed), and the column with large lateral teeth and a fimbriate, longer median tooth.

Most species in this genus were originally in the genus *Epidendrum* until Dressler moved them to *Encyclia* in 1961. Brieger worked with the group and included them in his resurrected *Hormidium*, but the group was not well defined nor formally presented. Higgins included them in his large genus *Prosthechea*, but we feel that the group of species are easily defined and significantly separate from *Anacheilium* and other genera of the *Prosthechea* group, based on the above characteristics, most significant being the column and lip configurations.

Panarica Withner & Harding, *gen. nov. Panarica genus novum Anacheilium Hoffmannsegg similis sed floribus resupinatis, labello trilobato plerumque, dentibus columnaribus inaequalibus, medidente columari fimbriato differt.* Type species: *Panarica prismatocarpa* (Reichenbach f.) Withner & Harding, *gen. nov.*

The genus name is a combination of Panama and Costa Rica, reflecting the center of distribution of the species contained in the genus.

Species of *Panarica*

Panarica brassavolae, 208
Panarica ionocentra, 211
Panarica mulasii, 212
Panarica neglecta, 212
Panarica prismatocarpa, 213
Panarica tardiflora, 214

Key to Species of *Panarica*

1a. Lip unlobed .... *Panarica brassavolae*
1b. Lip three-lobed .... go to 2
2a. Lip more that 4 cm long, sepals and petals unspotted .... *Panarica ionocentra*
2b. Lip less than 3.5 cm long, sepals and petals brown spotted .... go to 3
3a. Inflorescence on fully mature growth .... *Panarica tardiflora*
3b. Inflorescence on emerging new growth .... go to 4
4a. Sepals warty (short protuberances) on reverse .... *Panarica neglecta*
4b. Sepals smooth on reverse .... *Panarica prismatocarpa*

Panarica brassavolae
FIGURE 6-2, PLATE 64

Panarica brassavolae (Reichenbach f.) Withner & Harding, *comb. nov.* Basionym: *Epidendrum brassavolae* Reichenbach f. 1852. *Botanische Zeitung (Berlin)* 10: 729.

SYNONYMS

Encyclium brassavolae (Reichenbach f.) Lindley ex Stein. 1892. *Orchideenbuch* 224.

Encyclia brassavolae (Reichenbach f.) Dressler. 1961. *Brittonia* 13: 264.

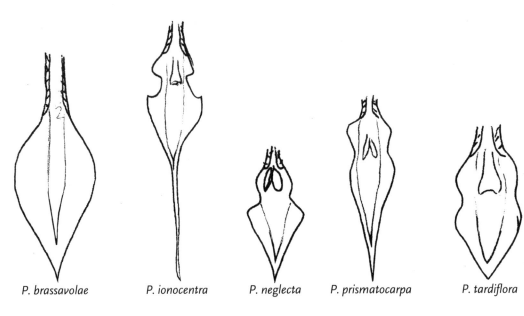

FIGURE 6-1. *Panarica* lips.

Hormidium brassavolae (Reichenbach f.) Brieger. 1977. Schlechter's *Die Orchideen*, ed. 3, p. 576.

Prosthechea brassavolae (Reichenbach f.) W. E. Higgins. 1997. *Phytologia* 82 (5): 381.

DERIVATION OF NAME

Honors Antonio Musa Brasavole, a fifteenth-century Italian physician specializing in medicinal plants.

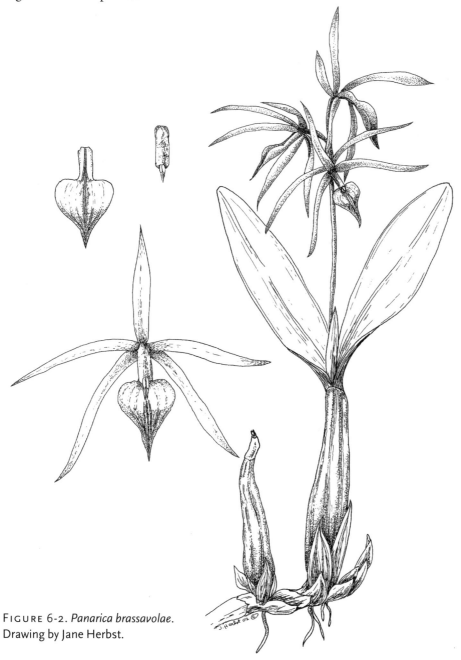

FIGURE 6-2. *Panarica brassavolae*. Drawing by Jane Herbst.

DESCRIPTION

Pseudobulbs loosely clustered to 4.5 cm apart, ovoid, somewhat flattened. Leaves two. Inflorescence simple, with three to fifteen flowers. Sepals and petals pale green, greenish yellow to olive tan, lip basally cream, distally violet purple. Lip unlobed, adnate to column for one-third the length of the column, the lip claw expanding into the blade at the apex of the column, the blade ovate, pointed, base of blade folded upward with fleshy margins, with a fleshy keel running the length of the blade. Column midtooth longer than lateral teeth. Capsule triangular in cross section.

HABITAT AND DISTRIBUTION

Central America and México. Found at 1200–2500 meters in elevation, in rather wet pine-oak and evergreen forest, sometimes on rocks.

FLOWERING TIME

January to September.

CULTURE

These plants are easy to grow and should be part of any mixed collection. They prefer intermediate temperatures, high humidity, and daily watering during the spring and summer with a decrease in watering in the fall and winter. Fertilizer should be given throughout the year. They grow well in baskets but will do well in pots or can be mounted if you can water them enough. They like high diffuse light and our plants in the Pacific Northwest get nearly full sun for several hours a day in the summer.

COMMENT

The plant is striking. It is remarkable that for a showy species described in 1852 that there are not more synonyms for this plant. Obviously when it was described it was well done, so that future taxonomists would remember it and not err in redescribing it.

MEASUREMENTS

Pseudobulbs 9–18 cm long, 3–5 cm wide
Leaves 14–28.5 cm long, 3–5.5 cm wide
Inflorescence 13–40 cm long
Spathe 4.8–15 cm long
Sepals 36–55 mm long, 3.5–6 mm wide
Petals 32–47 mm long, 2–4 mm wide
Lip 30–43 mm long, 11.5–16 mm wide
Column 12–15 mm long

Panarica ionocentra
PLATE 65

Panarica ionocentra (Reichenbach f.) Withner & Harding, *comb. nov.* Basionym: *Epidendrum ionocentrum* Reichenbach f. 1888. *Gard. Chron.* 2: 8. Type: probably Costa Rica, *Endres & Pfau s.n.* (holotype: W).

SYNONYMS
Epidendrum auriculigerum Reichenbach f. 1888. *Gard. Chron.* 2: 34.
Epidendrum prismatocarpum Reichenbach f. var. *ionocentrum* (Reichenbach f.) Teuscher. 1969. *American Orchid Society Bulletin* 38: 398.
Hormidium auriculigerum (Reichenbach f.) Brieger. 1977. Schlechter's *Die Orchideen*, ed. 3, p. 575.
Hormidium ionocentrum (Reichenbach f.) Brieger. 1977. Schlechter's *Die Orchideen*, ed. 3, p. 576.
Encyclia ionocentra (Reichenbach f.) D. E. Mora-Retana & J. García-Castro. 1991. *Brenesia* 33: 124.
Prosthechea ionocentra (Reichenbach f.) W. E. Higgins. 1997. *Phytologia* 82 (5): 381.

DERIVATION OF NAME
Greek *iono*, "purple," and *centrum*, "center, midpoint," referring to the lip coloration.

DESCRIPTION
A short broad bulb bearing a raceme of twenty to twenty-four flowers. Sepals and petals lemon colored to greenish brown. Lip violet or purple with a white center.

HABITAT AND DISTRIBUTION
Costa Rica and Panama. Found at 900–1600 meters in elevation.

FLOWERING TIME
May to August.

CULTURE
This species requires intermediate conditions with watering frequently in spring and summer and less in fall and winter. High light encourages free blooming. The plants can be grown mounted or in pots.

COMMENT
Similar to *Panarica prismatocarpa* but lacking spots on sepals and petals. Flowers distinctly larger than those of *P. prismatocarpa* but the shape is the same.

Panarica mulasii

Panarica mulasii (Soto Arenas & L. Cervantes) Withner & Harding, *comb. nov.* Basionym: *Prosthechea mulasii* Soto Arenas & L. Cervantes. 2002. *Icones Orchidacearum* 5–6: plate 651.

HABITAT
Mexico

DESCRIPTION
A close relative of *Panarica brassavolae*, but with a shorter rhizome, one to two leaves per pseudobulb, pseudobulbs more ovoid and smaller, and the raceme has only four to eight flowers. The midtooth of the column widens at the halfway mark and then becomes narrow and pointed, where the column of *P. brassavolae* does not appear to widen midlength. The column and lip insertion are simpler than the U-shaped insertion characteristic of *P. brassavolae*.

Panarica neglecta

Panarica neglecta (Pupulin) Withner & Harding, *comb. nov.* Basionym: *Prosthechea neglecta* Pupulin. 2001. *Selbyana* 22: 21. Type: Costa Rica, San José, Dota, San Pedro, ca. 1900 meters, *M. Flores s.n.* (holotype: USJ).

DERIVATION OF NAME
Latin *neglectus*, "neglected," alluding to the long time the species was in cultivation without specific recognition.

DESCRIPTION
A lithophyte, with an elongated rhizome. Pseudobulbs narrowly ovate to linear-conic. Leaves two. Flowers eight to ten on a loose raceme. Sepals and petals yellowish cream spotted with purple. Sepals warty on reverse (outside or axial surface). Lip rose-purple, three-lobed; lateral lobes rounded, slightly reflexed; midlobe sagittate, concave, and pointed. Callus formed by two fleshy keels uniting centrally and diverging at apex.

HABITAT AND DISTRIBUTION
Costa Rica. Found at 1900–2800 meters in elevation, in wet montane forest.

FLOWERING TIME
March to June.

MEASUREMENTS

Pseudobulbs 26.5 cm long, 3.2–3.8 cm wide
Leaves 26–32 cm long, 3.2–3.8 cm wide
Inflorescence 33 cm long
Sepals 19 mm long, 4–4.5 mm wide
Petals 18 mm long, 3.5 mm wide
Lip 15 mm long, 6 mm wide
Column 8.5 mm long

Panarica prismatocarpa
PLATES 66, 67

Panarica prismatocarpa (Reichenbach f.) Withner & Harding, *comb. nov.* Basionym: *Epidendrum prismatocarpum* Reichenbach f. 1852. *Botanische Zeitung (Berlin)* 10: 729. Type: Panama. Volcano Chiriqui, November, 2400 meters, Warscewicz s.n. (W).

SYNONYMS

Epidendrum maculatum Hort. ex Reichenbach f. 1865. *Xenia Orchidacea* 2: 83, in syn.
Epidendrum uro-skinneri Hort. ex Reichenbach f. 1865. *Xenia Orchidacea* 2: 83, in syn.
Encyclia prismatocarpa (Reichenbach f.) Dressler. 1961. *Brittonia* 13 (3): 265.
Hormidium prismatocarpum (Reichenbach f.) Brieger. 1977. Schlechter's *Die Orchideen*, ed. 3, p. 574.
Prosthechea prismatocarpa (Reichenbach f.) W. E. Higgins. 1997. *Phytologia* 82 (5): 381.

DERIVATION OF NAME

Latin *prismaticus*, "prism-shaped," and *carpus*, "fruit."

DESCRIPTION

A clumping herb. Pseudobulbs cylindric, ovoid. Leaves two or three, somewhat separated on apex of bulb. Inflorescence erect, few- to many-flowered raceme, the flowers resupinate. Sepals and petals greenish white with purple spots. Lip greenish white with three short purple stripes. Lip three-lobed, free from column nearly to the base, subsagittate (shaped like arrowhead with lobes pointing backward); midlobe large, shaped like a spade trowel, pointed. Callus extends from the base to nearly the apex of the lip. Column club-shaped, three-toothed at the apex.

HABITAT AND DISTRIBUTION
Central America. Found at 1200–3300 meters in elevation.

FLOWERING TIME
April to August.

CULTURE
Culture for these plants requires intermediate conditions, with frequent watering in spring and summer and less in fall and winter. High light is preferred for blooming. Plants can grow mounted or in pots. Eric Christenson says they do "really better in pots in Florida." This is not what we have found in our conditions in the Pacific Northwest, but we have more humidity and less heat throughout the year, which may explain the difference.

COMMENT
This is the type species for the genus *Panarica*.

MEASUREMENTS
Pseudobulbs 15 cm long, 5 cm wide
Leaves 12–33 cm long, 2–6 cm wide
Inflorescence 30 cm long
Spathe 10 cm long
Sepals 22–44 mm long, 3–5 mm wide
Petals 20–28 mm long, 3–5 mm wide
Lip 18–25 mm long, 7–9 mm wide
Column 10 mm long

Panarica tardiflora

Panarica tardiflora (D. E. Mora-Retana ex Pupulin) Withner & Harding, *comb. nov.* Basionym: *Prosthechea tardiflora* D. E. Mora-Retana ex Pupulin. 2002. *Lankesteriana* 3: 23. Type: Costa Rica, Guanacaste, Santa Cruz, between Juan Díaz and Vista al Mar, 500–600 meters, collected by Raúl Cascante Arias, 1992, flowered in cultivation at Tambor de Alajuela, 15 January, *F. Pupulin 2806* (holotype: USJ).

DERIVATION OF NAME
Latin *tardiflorus*, "late flowering," referring to the plant's habit of flowering on last year's growth.

DESCRIPTION

A clump-forming epiphyte. Rhizome short, covered with brown bracts. Pseudobulbs ovate. Flowers emerge from last year's pseudobulb. Sepals and petals yellowish cream to ochre spotted with purple. Lip white with three rose-purple blotches.

HABITAT AND DISTRIBUTION

Costa Rica. Found at 200–800 meters in elevation, in rain forests and seasonally wet forests.

FLOWERING TIME

January to April.

MEASUREMENTS

Pseudobulbs 6.2–8 cm long, 3.4–5 cm wide
Leaves 15–18 cm long, 3–4 cm wide
Inflorescence 27 cm long
Spathe 8 cm long
Sepals 22–23 mm long, 5–6.5 mm wide
Petals 22 mm long, 3.5 mm wide
Lip 20 mm long, 9.5 mm wide
Column 13 mm long

CHAPTER 7

The Genus *Pollardia*

We have worked with this group, thinking that their generic name should be *Epicladium*, meaning "upon a branch," thus being able to resurrect the name of Lindley's subgroup of *Epidendrum*. We have discovered that we cannot use this generic name as the type species has been transferred to another genus (by us) and the original genus description has no other species that could be used. So you say, "What are you talking about?" Read on and we'll give you a lesson in the problems of taxonomy.

When Lindley first defined *Epidendrum* subgenus *Epicladium*, he included such miscellaneous species as *Epidendrum aurantiacum* (synonym *Cattleya aurantiaca*), and *E. boothianum* (synonym *Prosthechea boothianum*). The key characteristics for the group were that the lip was nearly free from the column and that the flowers came from a sheath. These characteristics are not definitive enough to circumscribe an orchid taxon, and the name fell into disuse. Acuña used it for *Epicladium boothianum* and *Epicladium boothianum* var. *erythronioides*, but these species are today placed in other genera.

John K. Small raised this group of plants to the generic level in 1913 when he published *The Flora of Miami*. Small presented the following definition of *Epicladium*, using *E. boothianum* as the type species:

> Epiphytic herbs with short flattened pseudobulbs which bear several short leaves and a relatively short mostly unbranched flowering stem which is subtended by a long foliaceous spathe. Flowers erect or ascending, mostly showy, subtended by bracts. Lateral sepals rather short, about as wide as the dorsal one. Petals nearly resembling the lateral sepals or dilated toward the apex. Lip shorter than the sepals and petals, the blade usually rhombic. Column partly adnate to the lip. Capsules nodding and winged.

We initially decided to retain *Epicladium* for a number of species in our complex from *Epidendrum* today, applying it in its original sense, knowing that these plants didn't really fit the type description. It did not really accomplish the task of using Lindley's *Epicladium* in that the lips were not free of the column, as he defined them. Surprising what you can overlook in convenience when you want something to fit. However, remember that the type species for the genus *Epicladium* is *E. boothianum*. We have determined this species fits more correctly into the genus *Prosthechea*, which leaves the genus *Epicladium* as synonym, being absorbed, so to speak, within the genus *Prosthechea*. The only way to restore *Epicladium* at this time would be to have a group of species similar to *E. boothianum* elevated to their own genus, which upon reflection we cannot justify.

So now we have a genus with no name and no type species. There also are no other names that take precedent in this group of species. This group started out in *Epidendrum*, was then moved to *Encyclia*, a few members were moved into *Anacheilium* and *Hormidium*, and most recently they have been placed in *Prosthechea*. Since the defining characteristics of *Prosthechea* exclude these species, we have to come up with another name for the genus. We have decided to honor Glenn E. Pollard for his extensive work with Mexican orchids by creating the genus *Pollardia*. It was with his book, coauthored by Robert Dressler, *The Genus Encyclia in México* (1974), that we were able to find so much of the information included in this and other chapters.

Pollardia Withner & Harding, *gen. nov. Pollardia genus novum Prosthechea Knowles et Westcott similis sed dentibus columnaribus aequalibus, inflorescentia spathacea differt.* Type species: *Pollardia livida* (Lindley) Withner & Harding.

We define this genus as having (1) a lip partially attached to the column; (2) resupinate flowers that come from sheath; and (3) column teeth of equal or subequal length and generally of the same size. The resupinate flowers separate this genus from *Anacheilium* and the equal column teeth separate it from *Prosthechea*. The plants are generally small, less than 20 cm tall, though a few species do get large. The pseudobulbs are spindle-shaped, slightly flattened; and if there is a thin sheath covering the bulb it only covers it partway. The flowers are colorful and pleasingly shaped although small, held well above the foliage, and generally have showy lips. As the plants are small and compact, many of them could be part of any collection.

Culture for this group differs from *Anacheilium* in that the plants generally require much less water and have definite rest periods, some species requiring a very strict dry rest period. They grow well either mounted or in pots that drain quickly with drying in between waterings. They are not heavy feeders, and fertilizer type can be just what you are giving other plants. They can handle rather bright light and some actually prefer full light. These plants generally grow in clumps, or if they do spread out, the distance between pseudobulbs is short so they still stay in their container or on their mount for some time. The flowers are held well above or apart from the foliage so that although the flowers may be small by some standards a well-bloomed plant will show off well.

Species of *Pollardia*

Pollardia campylostalix, 220
Pollardia concolor, 222
Pollardia ghiesbreghtiana, 224
Pollardia greenwoodiana, 226
Pollardia hastata, 228
Pollardia linkiana, 229
Pollardia livida, 231
Pollardia michuacana, 233
Pollardia obpiribulbon, 235
Pollardia pringlei, 237
Pollardia pterocarpa, 238
Pollardia punctulata, 240
Pollardia semiaptera, 242
Pollardia tripunctata, 244
Pollardia varicosa, 245
Pollardia venosa, 247

Key to Species of *Pollardia*

1a. Lip unlobed . . . . go to 2
1b. Lip three-lobed . . . . go to 3
2a. Lip notched, column very thick, more than half as wide as long . . . . *Pollardia ghiesbreghtiana*
2b. Lip with sharp tip, column slender . . . . *Pollardia pringlei*
3a. Lip with multiple prominent warts . . . . go to 4
3b. Lip with no warts . . . . go to 5
4a. Lip strongly plicate-undulate . . . . *Pollardia livida*
4b. Lip edges smooth . . . . *Pollardia varicosa*
5a. Lip without prominent keels . . . . go to 17
5b. Lip with keels . . . . go to 6
6a. Lip one color or with a few spots . . . . go to 7
6b. Lip with veining of different color . . . . go to 12
7a. Lip notched at apex . . . . *Pollardia michuacana*
7b. Lip unnotched at apex . . . . go to 8
8a. Lateral lobes barely more than acute angles on sides of lip, sometimes absent . . . . go to 9
8b. Lateral lobes prominent . . . . go to 11
9a. Lip blade subquadrate, square or shield-shaped, wider than long . . . . *Pollardia hastata*
9b. Lip longer than wide . . . . go to 10
10a. Column blue-purple, straight . . . . *Pollardia tripunctata*
10b. Column green with reddish purple spots, up-curved . . . . *Pollardia campylostalix*

11a. Midlobe of lip transversely oblong, stiff pointed .... *Pollardia concolor*
11b. Midlobe of lip rhomboid, point less acute, not stiff........... *Pollardia punctulata*
12a. Lip with peach to red-brown veining .... go to 13
12b. Lip with wine-colored veining... go to 14
13a. Lateral lobes of lip larger than the bluntly pointed midlobe ... *Pollardia pterocarpa*
13b. Lateral lobes of lip about same size as the rounded midlobe.....................
.... *Pollardia obpiribulbon*
14a. Lateral lobes when spread extend wider than midlobe........... *Pollardia linkiana*
14b. Lateral lobes when spread do not extend as wide as the midlobe......... go to 15
15a. Lateral lobes of the lip with purple veining .... *Pollardia venosa*
15b. Lateral lobes of lip with no contrasting veining .... go to 16
16a. Lip forward projecting .... *Pollardia semiaptera*
16b. Lip somewhat reflexed .... *Pollardia greenwoodiana*

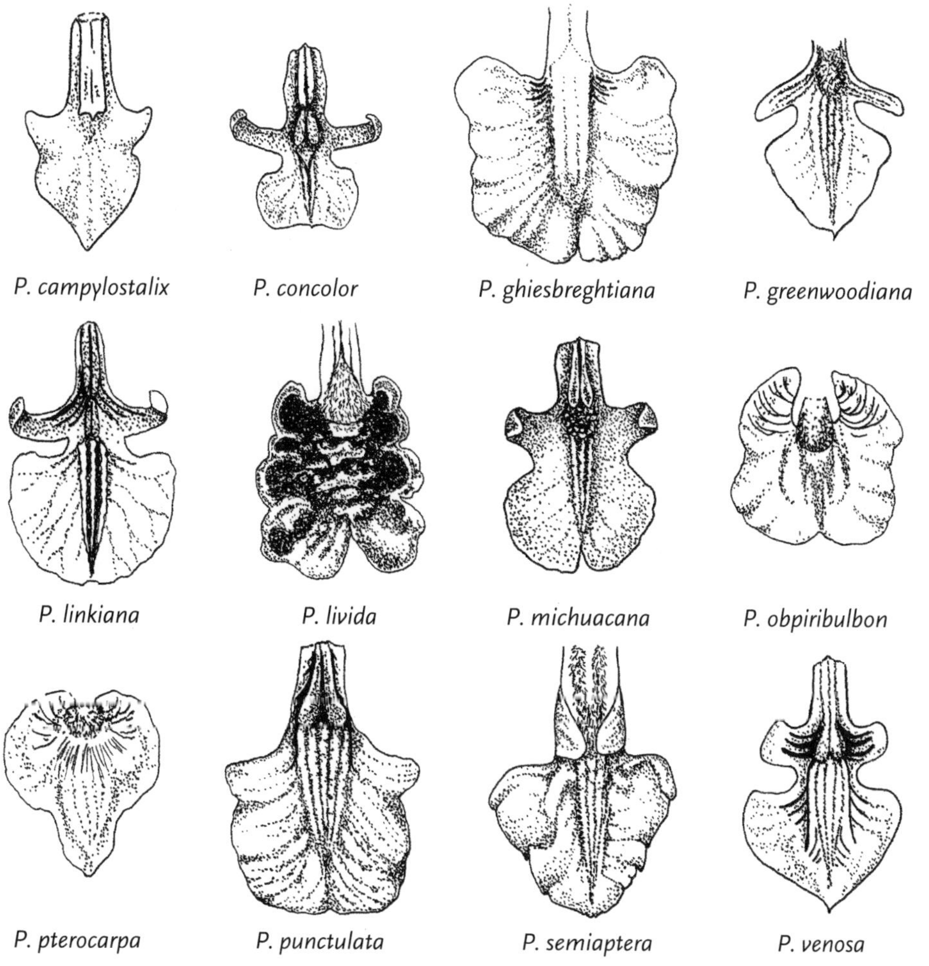

P. campylostalix   *P. concolor*   *P. ghiesbreghtiana*   *P. greenwoodiana*

P. linkiana   *P. livida*   *P. michuacana*   *P. obpiribulbon*

P. pterocarpa   *P. punctulata*   *P. semiaptera*   *P. venosa*

FIGURE 7-1. *Pollardia* lips.

Pollardia campylostalix

FIGURE 7-2, PLATE 68

Pollardia campylostalix (Reichenbach f.) Withner & Harding, *comb. nov.* Basionym: *Epidendrum campylostalix* Reichenbach f. 1852. *Botanische Zeitung (Berlin)* 10: 730.

SYNONYMS

Encyclia campylostalix (Reichenbach f.) Schlechter. 1922. *Repertorium Specierum Novarum Regni Vegetabilis, Beihefte* 17: 45.
Prosthechea campylostalix (Reichenbach f.) W. E. Higgins. 1997. *Phytologia* 82 (5): 381.

DERIVATION OF NAME

Greek *campylo*, "bent," and *stalix*, "stake," referring to the curved column.

DESCRIPTION

Plants stout to 40 cm tall. Pseudobulbs ovoid to ellipsoid, strongly compressed. Leaf one, gray-green (glaucous) especially on lower surface. Inflorescence a loosely flowered raceme or panicle with several branches. Flowers nodding and spreading, resupinate. Sepals and petals similar, narrowly lanceolate, acuminate, grayish green on outer surface, dull wine within. Lip obscurely or deeply three-lobed, from a long narrow claw, united with the column at base, white with reddish spots near base, lateral lobes short and rounded or oblong and obtuse with deep sinus separating them from midlobe, midlobe more or less orbicular, rounded or subapiculate at apex, occasionally separated from the lateral lobes by a broad isthmus; disk with fleshy callus on the claw just below lateral lobes giving rise to three small keels that extend onto the midlobe. Column green and reddish purple marked with red spots, three-lobulate at apex with an upward curve to column. Ovary glaucous, three-angled with wings.

HABITAT AND DISTRIBUTION

Costa Rica, Guatemala, and Panama. Found at 1800–2000 meters in elevation, epiphytic on trees.

CULTURE

Culture for this species should be intermediate temperatures, bright diffuse light, and watering in active growing season, with less water after the pseudobulbs mature.

MEASUREMENTS
Pseudobulbs 3.5–10.5 cm long, 3–4 cm wide
Leaves 9–30 cm long, 2.5–8 cm wide
Inflorescence 10–38 cm long
Spathe 3–6 cm long
Sepals 12–25 mm long, 15–35 mm wide
Petals 12–25 mm long, 15–35 mm wide
Lip 12–20 mm long
Lateral lobes 1–3 mm wide
Midlobe 4.5–6 mm wide
Column 9 mm long

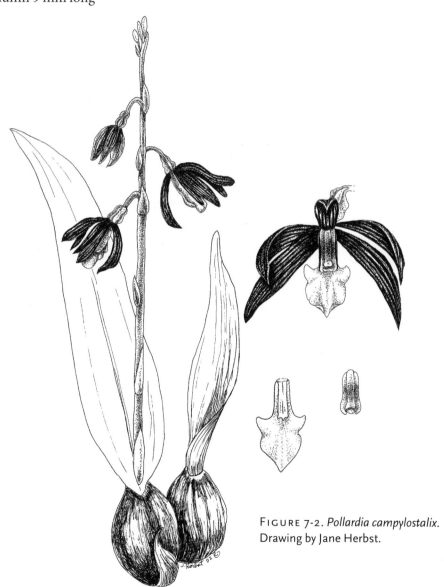

FIGURE 7-2. *Pollardia campylostalix*. Drawing by Jane Herbst.

Pollardia concolor
FIGURE 7-3, PLATE 69

Pollardia concolor (Llave & Lexarza) Withner & Harding, *comb. nov.* Basionym: *Epidendrum concolor* Llave & Lexarza. 1825. *Novorum Vegetabilium Descriptiones* 2: 25.

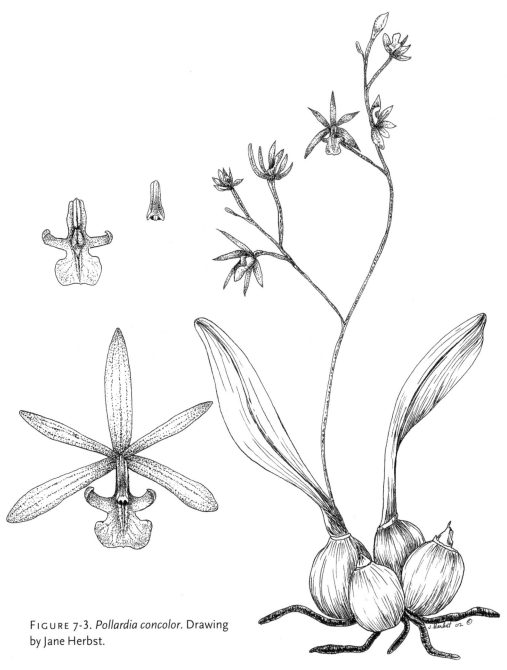

FIGURE 7-3. *Pollardia concolor*. Drawing by Jane Herbst.

SYNONYMS

Epidendrum pruinosum A. Richard & Galeotti. 1845. *Ann Sci. Nat*, ser. 3, 3: 20.

Epidendrum amabile Linden & Reichenbach f. 1855. *Bonplandia* 3: 219.

Encyclia concolor (Llave & Lexarza) Schlechter. 1918. *Beih. Bot. Centralbl.* 36 (2): 472.

Encyclia pruinosa (A. Richard & Galeotti) Schlechter. 1918. *Beih. Bot. Centralbl.* 36 (2): 473.

Encyclia amabilis (Linden & Reichenbach f.) Schlechter. 1918. *Beih. Bot. Centralbl.* 36 (2): 473.

Prosthechea concolor (Llave & Lexarza) W. E. Higgins. 1997. *Phytologia* 82 (5): 381.

DERIVATION OF NAME

Latin *concolor*, "uniform color," referring to the color of the lip.

DESCRIPTION

Pseudobulbs clustered, ovoid, slightly flattened. Leaves one or two. Inflorescence simple or branched with three to thirty flowers. Sepals and petals greenish brown, brownish, yellow, or buff. Lip white, basally adnate to column, three-lobed (rarely entire without lateral lobes), lateral lobes ovate, midlobe transversely oblong, stiff pointed. Callus oblong, fleshy, becoming three thick veins that nearly extend to margin of lip. Column midtooth obtuse, fleshy, subequal to lateral teeth. Capsule triangular in cross section.

HABITAT AND DISTRIBUTION

México. Found at 1500–2100 meters in elevation, in oak and oak-pine forests.

FLOWERING TIME

December to June.

CULTURE

Culture for this plant should be warm temperatures with high light. Attention to watering is important, as in habitat the plant normally has a very dry rest period.

COMMENT

This species is very similar to *Pollardia michuacana* but *P. concolor* is a much smaller plant and its callus and lip are fleshy. *Pollardia michuacana* has pseudobulbs more than 5 cm long while those of *P. concolor* are less than 4 cm long.

MEASUREMENTS

Pseudobulbs 1.2–4 cm long, 1.5–3 cm wide
Leaves 6–18 cm long, 1–3 cm wide
Inflorescence 12–45 cm long

Spathe 4.5 cm long
Sepals 8–15 mm long, 2–4.5 mm wide
Petals 7–12 mm long, 1–3.5 mm wide
Lip 10–12 mm long, 4–7 mm wide
Column 4–5 mm long

Pollardia ghiesbreghtiana
FIGURE 7-4, PLATE 70

Pollardia ghiesbreghtiana (A. Richard & Galeotti) Withner & Harding, *comb. nov.* Basionym: *Epidendrum ghiesbreghtianum* A. Richard & Galeotti. 1845. *Annales des Sciences Naturelles; Botanique,* sér. 3, 3: 19.

SYNONYMS

Encyclia ghiesbreghtiana (A. Richard & Galeotti) Dressler. 1961. *Brittonia* 13 (3): 264.

Hormidium ghiesbreghtianum (A. Richard & Galeotti) Brieger. 1977. Schlechter's *Die Orchideen,* ed. 3, p. 574.

Prosthechea ghiesbreghtiana (A. Richard & Galeotti) W. E. Higgins. 1997. *Phytologia* 82 (5): 381.

DERIVATION OF NAME

Honors Auguste Boniface Ghiesgbreght, a famous nineteenth-century collector of plants in México.

DESCRIPTION

Pseudobulbs clustered, ovoid. Leaves two or three. Inflorescence of one to three resupinate flowers. Sepals and petals pale green, densely spotted or streaked with brown (also described as maroon with narrow greenish white margins, reverse side olivaceous). Lip white with few short purple veins near base, column green with brown spots. Lip unlobed to three-lobed (if lateral lobes present then small), basally adnate to column, blade square (subquadrate-orbicular), base wedge-shaped, apex blunt with central notch. Callus oblong, becoming three fleshy veins that run to near the apical margin. Column teeth obtuse, subequal. Capsule three-winged.

HABITAT AND DISTRIBUTION

México. Found at 2000–2700 meters in elevation, in pine-oak forests, seasonally rather wet.

FLOWERING TIME

February to May.

CULTURE

Culture is intermediate to warm temperatures with a pronounced seasonal (winter) dry period.

COMMENT

Patricia's plant has tepals that were pale green with very little brown. The species is similar to *Pollardia hastata* but *P. hastata* is smaller and has pseudobulbs with only one leaf.

FIGURE 7-4. *Pollardia ghiesbreghtiana*. Drawing by Jane Herbst.

MEASUREMENTS
Pseudobulbs 2–6.5 cm long, 0.5–2.4 cm wide
Leaves 6–20 cm long, 0.7–1.3 cm wide
Inflorescence 4–12 cm long
Spathe 4 cm long
Sepals 15–24 mm long, 3–8.5 mm wide
Petals 14–22 mm long, 3–6 mm wide
Lip 16–27 mm long, 16–30 mm wide
Column 5–6.5 mm long

Pollardia greenwoodiana
FIGURE 7-5

Pollardia greenwoodiana (Aquirre-Olavarrieta) Withner & Harding, *comb. nov.* Basionym: *Encyclia greenwoodiana* Aquirre-Olavarrieta. 1992. *Orquídea (México)* 12 (2): 205. Type: México, Oaxaca: 164 km of the Oaxaca road, PTO, concealed, 2020 meters, midportion of mountain forest of *Chiranthodendron and oak*, April 1988, specimen from cultivated material, 28 June 1989, *M. Soto Arenas 4106* (holotype: AMO; isotype: MEXU).

SYNONYM
Prosthechea greenwoodiana (Aquirre-Olavarrieta) W. E. Higgins. 1997. *Phytologia* 82 (5): 381.

DERIVATION OF NAME
Honors Ed Greenwood, a tireless student of México's rich orchid heritage.

DESCRIPTION
An epiphyte. Pseudobulbs reverse pear-shaped to elliptic. Flowers six to eight on inflorescence, opening successively. Sepals and petals yellow-green, lip white or cream. Lip three-lobed, lateral lobes falcate (scythe-shaped). Column apically three-toothed. Ovary green, triangular in cross section.

HABITAT AND DISTRIBUTION
México (Oaxaca), endemic in a narrow range or area. Found in montane forest.

FLOWERING TIME
Any season.

COMMENT
The species is similar to *Pollardia semiaptera* except the pseudobulbs are differ-

ently shaped, and the lateral lobes of the lip are narrower and curved differently with respect to the column.

MEASUREMENTS

Pseudobulbs 6–11 cm long, 1.6–2.8 cm wide
Leaves 12–18 cm long, 1.2–2 cm wide
Inflorescence 8.5–20 cm long
Spathe 2.5 cm long
Sepals 16–18 mm long, 3–3.5 mm wide
Petals 14–15 mm long, 2–3 mm wide
Lip 15 mm long, 6–8 mm wide
Column 7–8 mm long

FIGURE 7-5. *Pollardia greenwoodiana*. Drawing by Jane Herbst.

Pollardia hastata
PLATE 71

Pollardia hastata (Lindley) Withner & Harding, *comb. nov.* Basionym: *Epidendrum hastatum* Lindley. 1840. *Journal of Botany* 3: 82.

SYNONYMS
Encyclia hastata (Lindley) Dressler & G. E. Pollard. 1971. *Phytologia* 21 (7): 437.
Hormidium hastatum (Lindley) Brieger. 1977. Schlechter's *Die Orchideen*, ed. 3, p. 574.
Prosthechea hastata (Lindley) W. E. Higgins. 1997. *Phytologia* 82 (5): 381.

DERIVATION OF NAME
Latin *hastatus*, "abruptly enlarged at the base into two acute diverging lobes (two triangles pointing outward), like the head of a halberd, an axe with two protrusions at base."

DESCRIPTION
Pseudobulbs clustered, ellipsoid-ovoid, slightly flattened. Leaf one. Inflorescence unbranched, with one to eight flowers. Sepals and petals purplish brown with darker veins, lip white. Lip three-lobed, adnate to column for about half its length, lateral lobes scarcely more than prominent angles at the base of the lip, midlobe of lip subquadrate, square or shield-shaped, obtuse, widest near base, lateral lobes small. Callus quadrate running to three fleshy veins, which run nearly to the apical margin. Column teeth obtuse, subequal. Capsule triangular in cross section.

HABITAT AND DISTRIBUTION
México. Found at 2400–2700 meters in elevation, in pine-oak forests.

FLOWERING TIME
April to May.

COMMENT
The species resembles *Pollardia ghiesbreghtiana*, which is larger and has a short thick column and two pseudobulb leaves. It also resembles *P. pringlei*, which is smaller, more delicate and has a lip that is wider than long.

MEASUREMENTS
Pseudobulbs 3.5–4.5 cm long, 1.8–2 cm wide
Leaves 9–16 cm long, 0.9–1.8 cm wide
Inflorescence 10–25 cm long

Spathe 1.5–2 cm long
Sepals 12–18 mm long, 2.5–6 mm wide
Petals 11–15 mm long, 1.5–3.5 mm wide
Lip 8–15 mm long, 10–17 mm wide
Column 6 mm long

Pollardia linkiana
FIGURE 7-6, PLATE 72

Pollardia linkiana (Klotzsch) Withner & Harding, *comb. nov.* Basionym: *Epidendrum linkianum* Klotzsch. 1838. *Allgemeine Gartenzeitung* 6: 299.

SYNONYMS

Epidendrum pastoris Link & Otto. 1828. *Icon. Pl. Rar.* 23, t. 12. (non *E. pastoris* Llave & Lexarza).
Encyclia linkiana (Klotzsch) Schlechter. 1918. *Beih. Bot. Centralbl.* 36 (2): 472.
Prosthechea linkiana (Klotzsch) W. E. Higgins. 1997. *Phytologia* 82 (5): 381.

DERIVATION OF NAME

Honors Johann Heinrich Friedrich Link (1767–1851), botanist and founder of the Berlin herbarium.

DESCRIPTION

Pseudobulbs separated by 2–2.5 cm along the rhizome, ellipsoid or spindle-shaped, somewhat flattened. Leaves one to four, long and narrow. Flowers five to twelve. Sepals and petals green or yellow-green, more or less heavily lined and suffused with red-brown or purplish brown, midlobe of lip cream, lateral lobes yellow with red violet veins. Lip three-lobed, adnate to column for half the length of the column, the lateral lobes oblong, when spread the lateral lobes are together wider than the midlobe, the midlobe suborbicular with fleshy veins. Callus obovate, pubescent, becoming fleshy veins in front. Column green, suffused with purple-brown, with three teeth, midtooth surpassing the lateral teeth. Capsule three-winged.

HABITAT AND DISTRIBUTION

México. Found at 500–2300 meters in various forest types.

FLOWERING TIME

Recorded from December to August but probably throughout the year.

MEASUREMENTS
Pseudobulbs 3.5–8.5 cm long, 0.6–2.6 cm wide
Leaves 11–22 cm long, 0.8–1.8 cm wide
Inflorescence 10–15 cm long
Sepals 7–15 mm long, 2.5–5 mm wide
Petals 9–18 mm long, 2–4 mm wide
Lip 9–10.5 mm long, 5–6 mm wide
Column 5–7 mm long

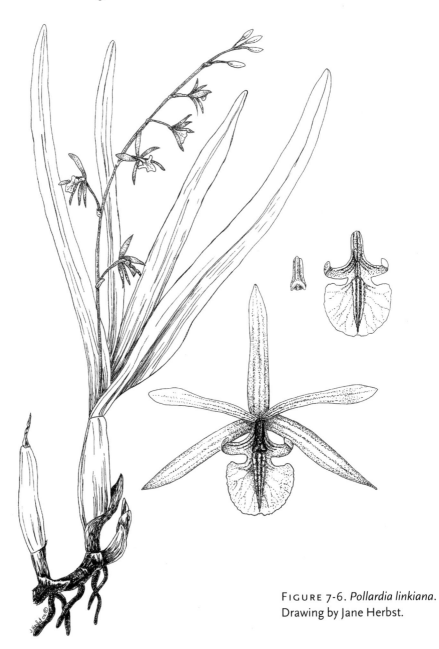

FIGURE 7-6. *Pollardia linkiana*.
Drawing by Jane Herbst.

Pollardia livida
FIGURE 7-7, PLATES 73, 74

Pollardia livida (Lindley) Withner & Harding, *comb. nov.* Basionym: *Epidendrum lividum* Lindley. 1838. *Edward's Botanical Register* 24, misc. 51.

SYNONYMS

Epidendrum tessellatum Bateman *ex* Lindley. 1838. *Edward's Botanical Register* 24, misc. 9; H, Bot. Mag., t. 3638.

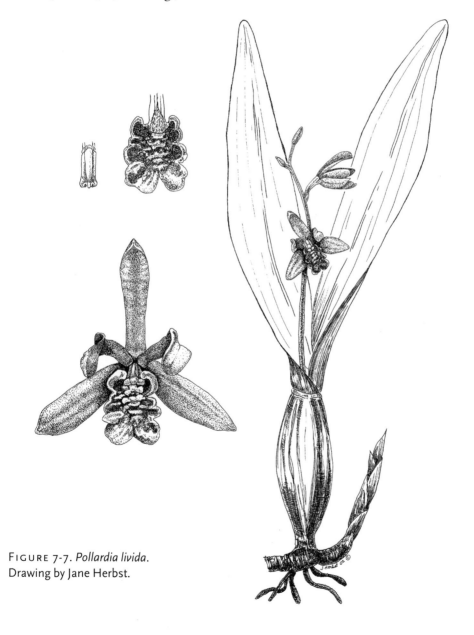

FIGURE 7-7. *Pollardia livida*. Drawing by Jane Herbst.

Epidendrum articulatum Klotzsch. 1838. *Allgemeine Gartenzeitung* 6: 297.

Epidendrum condylochilum Lehmann & Kraenzlin. 1899. *Engl. Bot. Jahrb. Syst.* 26: 459.

Epidendrum henrici Schlechter. 1906. *Repertorium Specierum Novarum Regni Vegetabilis.* 3: 108.

Epidendrum deamii Schlechter. 1918. *Beih. Bot. Centralbl.* 36, pt. 2: 402.

Encyclia tessellata Schlechter. 1918. *Beih. Bot. Centralbl.* 36, pt. 2: 474.

Epidendrum dasytaenia Schlechter. 1921. *Repertorium Specierum Novarum Regni Vegetabilis, Beihefte* 8: 71.

Encyclia deamii (Schlechter) Hoehne. 1952. *Arquivos de botanica do estado de São Paulo,* n.s., 2: 151.

Encyclia livida (Lindley) Dressler. 1961. *Brittonia* 13: 264.

Hormidium lividum (Lindley) Brieger. 1977. Schlechter's *Die Orchideen,* ed. 3, p. 574.

Anacheilium lividum (Lindley) Pabst, Moutinho & A. V. Pinto. 1981. *Bradea* 3: 23.

Prosthechea livida (Lindley) W. E. Higgins. 1997. *Phytologia* 82 (5): 381.

DERIVATION OF NAME

Latin *lividus,* "effect of adding gray and black to the range of hues between blue and red, blue or leaden color, bluish appearance caused by a bruise, congestion of blood vessel."

DESCRIPTION

An epiphyte. Pseudobulbs pale to light green compressed, slightly wrinkled, rather soft, somewhat variable in shape but generally rather slender. Leaves two or three. Inflorescences unbranched to 15 cm. Flowers five to seven, resupinate. Sepals and petals pale gray-green on the outer face, the inner face rather brighter green with puce marks, apex of sepals thickened and pointed. Lip three-lobed, though may be variable in lobe definition. Lip creamy white with some purple veining. Margins of lip strongly wavy (undulate). Callus creamy white, densely hairy from base to midpoint, becoming three rows of fleshy, yellow-cream teeth (warts), reaching to near apex. Column triangular in cross section, green covered with dark brownish purple color, fleshy green-rose at apex, underside thinly covered with very fine short white hairs, three-toothed at apex. Anther white. Capsule three-winged.

HABITAT AND DISTRIBUTION

México to Peru. Found at 1000–1400 meters in semideciduous forest.

FLOWERING TIME

February to April; other sources say May to October.

CULTURE

Culture for this plant is seasonal, with daily watering in the spring and summer seasons of growth, and a dry rest when growth is complete. Temperatures should be warm to intermediate, and the light levels diffuse to bright.

COMMENT

Plates 73 and 74 show two color forms. Patricia has both forms in her collection and they bloom at the same time and other than the color and the size of the flower, they do appear to be the same. The orange form comes from México and the darker more livid colored form comes from Venezuela.

MEASUREMENTS

Pseudobulbs 3.5–8.5 cm long, 0.6–2.6 cm wide
Leaves 11–22 cm long, 0.8–1.8 cm wide
Inflorescence 10–15 cm long
Sepals 7–11 mm long, 2.5–5 mm wide
Petals 9–10 mm long, 2–4 mm wide
Lip 9–10.5 mm long, 5–6 mm wide
Column 5–5.5 mm long

Pollardia michuacana
FIGURE 7-8, PLATE 75

Pollardia michuacana (Llave & Lexarza) Withner & Harding, *comb. nov.* Basionym: *Epidendrum michuacanum* Llave & Lexarza. 1825. *Novorum Vegetabilium Descriptiones* 2: 26.

SYNONYMS

Epidendrum virgatum Lindley. 1840. *Journal of Botany* 3: 83.
Epidendrum virgatum var. *pallens* Reichenbach f. 1865. *Bonplandia* 4. 326.
    Dressler and Pollard (1974) question whether this variant actually is *P. michuacana*.
Encyclia virgata (Lindley) Schlechter. 1914. *Die Orchideen*, ed. 1, p. 212.
Encyclia michuacana (Llave & Lexarza) Schlechter. 1918. *Beih. Bot. Centralbl.* 36 (2): 472.
Encyclia amabilis Schlechter. 1918. *Beih. Bot. Centralbl.* 36 (2): 458.
Epidendrum icthyphyllum Ames. 1923. *Schedulae Orchidianae* 2: 28.

Encyclia icthyphylla (Ames) Hoehne. 1952. *Arquivos de botanica do estado de São Paulo*, n.s., 2: 152.

Prosthechea michuacana (Llave & Lexarza) W. E. Higgins. 1997. *Phytologia* 82 (5): 381.

DERIVATION OF NAME

After Michoacán state in México.

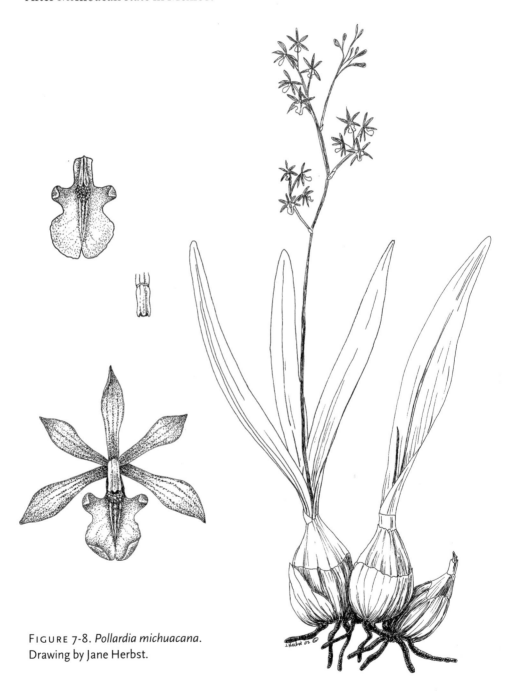

FIGURE 7-8. *Pollardia michuacana*. Drawing by Jane Herbst.

DESCRIPTION

Pseudobulbs clustered, pear-shaped to ovoid. Leaves two to four. Inflorescence branched, 40–200 cm long, with many resupinate flowers. Sepals and petals light to dark brown, lip cream or pale yellow, commonly speckled with minute purple dots. Lip three-lobed, adnate to column for basal third, lateral lobes oblong, obtuse, midlobe suborbicular. Callus thick, fleshy, sometimes with three veins on the base of the midlobe, lip otherwise smooth. Column green with three teeth, midtooth fleshy surpassing lateral teeth. Capsule triangular in cross section.

HABITAT AND DISTRIBUTION

Honduras, Guatemala, and México. Found at 1500–2800 meters, as a terrestrial in rather dry forest.

FLOWERING TIME

January to September.

CULTURE

Culture for this species requires an absolutely dry rest period, warm temperatures, and high light levels.

MEASUREMENTS

Pseudobulbs 5–12 cm long, 3–7 cm wide
Leaves 15–63 cm long, 2.5–6 cm wide
Inflorescence 40–200 cm long,
Sepals 10–15 mm long, 1.5–4.5 mm wide
Petals 9–13 mm long, 1–3 mm wide
Lip 10–11 mm long, 4–7 mm wide
Column 5–6.5 mm long

Pollardia obpiribulbon
FIGURE 7-9, PLATE 76

Pollardia obpiribulbon (Hágsater) Withner & Harding, *comb. nov.* Basionym: *Encyclia obpiribulbon* Hágsater. 1982. *Orquídea (México)* 8 (2): 386. Type: México, Oaxaca, Copala, Agua Fría, south end of Puerto Angel Road, 1372 meters, January 1980, *J. Pastrana & Hágsater 5745* (holotype: AMO; isotypes: AMES, ENCB, K, MEXU, MICH, SEL).

SYNONYM

Prosthechea obpiribulbon (Hágsater) W. E. Higgins. 1997. *Phytologia* 82 (5): 381.

DERIVATION OF NAME
Latin *pirum*, "pear," and *bulbus*, "bulb," referring to the shape of the pseudobulbs.

DESCRIPTION
An epiphyte, growing on oaks. Pseudobulbs obpyriform (inverted pear-shaped), the base being narrow forming a neck, widening below the middle, widest about the upper third of pseudobulb, slightly flattened. Inflorescence a raceme of five flowers, shorter or equal to the leaves. Sepals and petals brown with darker veins. Lip obscurely three-lobed, midlobe nearly as large as the lateral lobes combined, cream with reddish brown veins on the lateral lobes. Callus forming pubescent

FIGURE 7-9. *Pollardia obpiribulbon*.
Drawing by Jane Herbst.

crest in the middle of the lip blade. Column teeth red-brown, anther cream, ventral channel pubescent. Capsule three-winged.

HABITAT AND DISTRIBUTION

México in Oaxaca and Guerrero. Found at 1300–2100 meters in elevation, in mixed pine-oak forest and in treetops in evergreen forest.

FLOWERING TIME

October to March.

MEASUREMENTS

Pseudobulbs 10 cm long, 3 cm wide
Leaves 10–16 cm long, 8–18 cm wide
Inflorescence 4–13 cm long
Spathe 6–18 cm long
Sepals 10–15 mm long, 3.5 mm wide
Petals 8–12.5 mm long, 3.5 mm wide
Lip 8–11 mm long, 8–11 mm wide
Column 6 mm long

Pollardia pringlei
PLATE 77

Pollardia pringlei (Rolfe ex Ames) Withner & Harding, *comb. nov.* Basionym: *Epidendrum pringlei* Rolfe ex Ames. 1904. *Proceedings of the Biological Society of Washington* 17: 120.

SYNONYMS

Encyclia pringlei (Rolfe *ex* Ames) Schlechter. 1918. *Beih. Bot. Centralbl.* 36 (2): 473.
Hormidium pringlei (Rolfe ex Ames) Brieger. 1977. Schlechter's *Die Orchideen*, ed. 3, p. 574.
Prosthechea pringlei (Rolfe *ex* Ames) W. E. Higgins. 1997. *Phytologia* 82 (5): 381.

DERIVATION OF NAME

Honors Cyrus Guernsey Pringle, a famous nineteenth-century botanical collector of Mexican plants.

DESCRIPTION

Pseudobulbs clustered ovoid. Leaves one or two. Inflorescence unbranched with one to four resupinate flowers. Sepals and petals reflexed, pale green or pale brown, lip white with a few faint purple spots. Lip unlobed, adnate to column at base, oblong with a stiff sharp tip, callus oblong, becoming three fleshy veins which run to near the apical margin. Column dark purple, column teeth truncate, midtooth shorter than lateral teeth. Capsule triangular in cross section but not keeled or winged.

HABITAT AND DISTRIBUTION
México. Found at 1800–2500 meters, in rather wet pine-oak forests.

FLOWERING TIME
March to April.

COMMENT
The species resembles *Pollardia hastata* but the lip is wider and the flowers are smaller and more delicate.

MEASUREMENTS
Pseudobulbs 1.5–3.5 cm long, 0.4–1.2 cm wide
Leaves 4.5–9.5 cm long, 4–7.5 cm wide
Inflorescence 7–16 cm long
Sepals 9–15 mm long, 2–3.5 mm wide
Petals 8–13 mm long, 0.8–1.5 mm wide
Lip 7–10 mm long, 9–18 mm wide
Column 5–6 mm long

Pollardia pterocarpa
FIGURE 7-10, PLATE 78

Pollardia pterocarpa (Lindley) Withner & Harding, *comb. nov.* Basionym: *Epidendrum pterocarpum* Lindley. 1841. *Journal of Botany* 3: 82.

SYNONYMS
Epidendrum cinnamomeum A. Richard & Galeotti. 1845. *Ann. Sci. Nat.*, ser. 3, 3: 19.
Encyclia pterocarpa (Lindley) Dressler. 1961. *Brittonia* 13 (3): 265.
Prosthechea pterocarpa (Lindley) W. E. Higgins. 1997. *Phytologia* 82 (5): 381.

DERIVATION OF NAME
Greek *pter,* "winged," and *carpo,* "fruit," referring to the three-winged fruit.

DESCRIPTION
Pseudobulbs widely spaced, stalked, ovoid, somewhat flattened. Leaves two, sometimes three. Inflorescence unbranched with four to twelve resupinate flowers.

Sepals and petals pale green heavily lined with red-brown, lip cream or white, lateral lobes red or with red-violet lines. Lip weakly three-lobed, basally adnate to column, deltoid-ovate, cordate, broadly acute, lateral lobes curve upwards and basally clasp column, midlobe more or less recurved. Callus ovate, fleshy, truncate or three-toothed, pubescent, column teeth subequal. Capsule three-winged.

HABITAT AND DISTRIBUTION

Western México. Found at 900–2200 meters in elevation in oak and mixed forest, often growing on rocks.

FIGURE 7-10. *Pollardia pterocarpa*. Drawing by Jane Herbst.

FLOWERING TIME
Throughout year.

MEASUREMENTS
Pseudobulbs 3.5–16 cm long, 1–2.7 cm wide
Leaves 8.5–24 cm long, 1–2.7 cm wide
Inflorescence 7–30 cm long
Sepals 10–20 mm long, 2–4 mm wide
Petals 10–16 mm long, 2–3 mm wide
Lip 10–11 mm long, 8–9 mm wide
Column 6 mm long

Pollardia punctulata
FIGURE 7-11

Pollardia punctulata (Reichenbach f.) Withner & Harding, *comb. nov.* Basionym: *Epidendrum punctulatum* Reichenbach f. 1885. *Gard. Chron. n.s.* 24: 70. Type: Hort. Veitch (holotype: W [926: a lip and drawings]; illustration apparently based on the type, in *J. Day's Orchid Album* [27 May 1883], K).

SYNONYMS
Encyclia rhombilabia S. Rosillo. 1986. *Orquídea (México)* 10 (1): 145. Type: México, State of México, Santa Mónica, Municipality of Ocuilan, in oak forest at 2100 meters, epiphytic in association with *Encyclia pringlei*, 7 March 1981, *I. Aquirre & N. Pozos 13–211* (holotype: AMO; isotypes: AMES, IBUG, K, MEXU, SEL, US).
Prosthechea rhombilabia (S. Rosillo) W. E. Higgins. 1997. *Phytologia* 82 (5): 381.
Prosthechea punctulata (Reichenbach f.) Soto Arenas & Salazar. 2002. *Icones Orchidacearum* 5–6: plate 652.

DERIVATION OF NAME
Latin *punctatus*, "spotted."

DESCRIPTION
An epiphyte. Pseudobulbs ellipsoid, clustered, pale and whitish. Leaves one to three, whitish green. Inflorescence branched with twenty to seventy flowers. Sepals and petals green to dull brown to chestnut, lip white to yellow with or without purple blotch in central portion. Lip three-lobed united to base of column, lateral lobes half ovals, midlobe subrhomboid. Callus forming a fleshy mound with three keels extending to apex of the midlobe. Column semiquadrate with an apical tooth. Capsule three-sided. Fragrant in the morning of cinnamon.

HABITAT

México. Found at 1500–2400 meters in humid forest.

FLOWERING TIME

February to May.

COMMENT

This species is distinguished from *Pollardia concolor* and *P. michuacana* by having ovoid to ellipsoid pseudobulbs rather than pyriform or lenticular ones, both pseudobulbs and leaves pale glaucous green, a loosely branched inflorescence, and a rhomboid midlobe.

FIGURE 7-11. *Pollardia punctulata*. Drawing by Jane Herbst.

MEASUREMENTS
Pseudobulbs 4–11 cm long, 1.8–3.2 cm wide
Leaves 15–31 cm long, 2.2–3.1 cm wide
Inflorescence 60 cm long
Spathe 5 cm long
Sepals 14–16 mm long, 3–4 mm wide
Petals 14–15 mm long, 1.5–2 mm wide
Lip 13–15 mm long, 10 mm wide
Column 6–8 mm long

Pollardia semiaptera
FIGURE 7-12, PLATE 79

Pollardia semiaptera (Hágsater) Withner & Harding, *comb. nov.* Basionym: *Epidendrum tripterum* Lindley. 1840. *Hook. Jour. Bot.* 3: 83, non W. J. Hooker 1833— this species name not used because used for other plants. Lectotype: México, *Karwinski sub Zuccarini s.n.* (holotype: K; isotypes: M [×2]).

SYNONYMS
Encyclia semiaptera Hágsater. 1984. *Orquídea (México)* 9 (2): 234.
Prosthechea semiaptera (Hágsater) W. E. Higgins. 1997. *Phytologia* 82 (5): 381.

DERIVATION OF NAME
Semi, "half," and *aperature*, "open," referring to the flower that seems only partially open.

DESCRIPTION
An epiphyte or a lithophyte. Pseudobulbs clustered, ellipsoid, somewhat flattened. Leaves one to three. Flowers four to ten, opening successively. Sepals and petals yellowish, striped with diffused purplish brown lines on lower half of the interior surface. Lip white to yellowish, irregularly spotted purple-brown and blotched on lateral lobes and usually spotted on the midlobe. Column green basely, purplish to apex. Apical teeth whitish, anther yellow. Lip three-lobed, united to column for half the length of the column. Lateral lobes inrolled as to embrace column, but instead of forming tube, always slightly overlapping it. Callus rectangular formed by two parallel hairy ridges, apically becoming three irregular keels, the midkeel reaching the apex. Column with three prominent teeth, midtooth decurved. Capsule three-winged.

HABITAT AND DISTRIBUTION
México. Found at 500–2500 meters in pine oak forest.

MEASUREMENTS

Pseudobulbs 3–8 cm long, 0.6–2.2 cm wide
Leaves 10–21 cm long, 1.2–2.3 cm wide
Spathe 4 cm long
Sepals 12–17 mm long, 3–4 mm wide
Petals 11–13 mm long, 2.5–3 mm wide
Lip 12–14 mm long
Lateral and medial lobes 6–9 mm wide
Column 5–5.5 mm long

FIGURE 7-12. *Pollardia semiaptera*.
Drawing by Jane Herbst.

Pollardia tripunctata
FIGURE 7-13, PLATE 80

Pollardia tripunctata (Lindley) Withner & Harding, *comb. nov.* Basionym: *Epidendrum tripunctatum* Lindley. 1853. *Folia Orchidacea Epidendrum* 41.

SYNONYMS
Epidendrum micropus Reichenbach f. 1863. *Hamburger Garten- and Blumenzeitung* 19: 13.

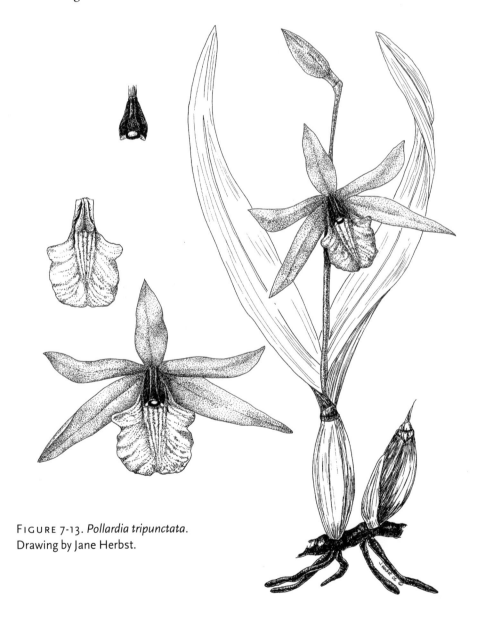

FIGURE 7-13. *Pollardia tripunctata*. Drawing by Jane Herbst.

Epidendrum diguetii Ames. 1922. *Schedulae Orchidianae* 1: 15.
Encyclia tripunctata (Lindley) Dressler. 1961. *Brittonia* 13 (3): 265.
Prosthechea tripunctata (Lindley) W. E. Higgins. 1997. *Phytologia* 82 (5): 381.

DERIVATION OF NAME
Latin *tri*, "three," and *punctata*, "spotted," referring to the contrastingly colored three dots (teeth) on the column.

DESCRIPTION
Pseudobulbs closely clustered, ovoid, slightly flattened, with whitish sheaths. Leaves two or three, deciduous. Inflorescence with one to five flowers. Sepals and petals pale green or yellowish green, lip white, column blue-purple, teeth at apex of column yellow. Lip adnate to about half the length of column, lateral lobes acute triangular toothlike, midlobe suborbicular. Callus oblong, pubescent, becoming three to five fleshy veins that run to near apex. Column with three apical teeth, midtooth longer. Capsule triangular in cross section. Fragrance heavenly, like cloves.

HABITAT AND DISTRIBUTION
México. Found at 1200–2000 meters in elevation.

FLOWERING TIME
March to June.

MEASUREMENTS
Pseudobulbs 2–4 cm long, 1.2–2 cm wide
Leaves 8–19 cm long, 0.9–1.5 cm wide
Inflorescence 3–10 cm long
Sepals 14–20 mm long, 2.5–4.5 mm wide
Petals 13–17 mm long, 1–3 mm wide
Lip 14–18 mm long
Column 7.5 mm long

Pollardia varicosa
PLATE 81

Pollardia varicosa (Lindley) Withner & Harding, *comb. nov.* Basionym: *Epidendrum varicosum* Bateman ex Lindley. 1838. *Edward's Botanical Register* 24, misc. 30. Type: Guatemala, *Skinner s.n.* (holotype: K).

SYNONYMS

Epidendrum leiobulbon (W. J. Hooker) W. J. Hooker. 1841. *Journ. Bot.* 3: 308, t. 10. *Epidendrum quadratum* Klotzsch. 1850. *Allgemeine Gartenzeitung* 18: 402.

Epidendrum chiriquense Reichenbach f. 1852. *Botanische Zeitung (Berlin)* 10: 730.

Epidendrum phymatoglossum Reichenbach f. 1852. *Botanische Zeitung (Berlin)* 10: 731.

Epidendrum lunaeanum A. Richard *ex* Lindley. 1853. *Folia Orchidacea Epidendrum* 23.

Encyclia chiriquensis (Reichenbach f.) Schlechter. 1918. *Beih. Bot. Centralbl.* 36, Abt. 2: 472.

Encyclia varicosa (Lindley) Schlechter. 1918. *Beih. Bot. Centralbl.* 36 (2): 472.

Epidendrum ramirezzi Gajón Sánchez, Mejores. 1930. *Orquídea (México)* 46, fig. 16.

Encyclia varicosa subsp. *leiobulbon* (W. J. Hooker) Dressler & Pollard. 1971. *Phytologia* 21 (7): 438.

Prosthechea varicosa (Lindley) W. E. Higgins. 1997. *Phytologia* 82: 381.

DERIVATION OF NAME

Latin *varicosus*, "abnormally enlarged in places."

DESCRIPTION

Plant slender. Pseudobulbs ovoid, spindle-shaped at base with long neck above. Leaves two or three. Raceme loose with few to many flowers. Flowers four to twenty-five, resupinate, fragrant. Sepals and petals green brown. Lip yellowish white, often purple spotted, deeply three-lobed, adnate to base of column with prominent claw, midlobe separated from lateral lobes by short isthmus. Callus thick, weakly three-toothed distally, sparsely pubescent, midlobe with three low crenulate veins (edged with rounded teeth that point forward) in center and smaller veins distally. Column stout, three-toothed, central tooth longer than lateral teeth, column usually purple blotched. Capsule three-winged.

HABITAT AND DISTRIBUTION

México to Panama. Found at elevations up to 3000 meters, terrestrial or on rocks, occasionally epiphytic in open woods.

COMMENT

Plant known to be extremely variable in length of the caulescent pseudobulb neck, the development of tubercles or warts on lip, and the size and shape of leaves.

MEASUREMENTS

Pseudobulbs 3–6 cm long, 2–4 cm wide
Leaves 8.5–25 cm long, 1.2–4.5 cm wide
Inflorescence 15–45 cm long
Spathe 1.5–5.5 cm long
Sepals 10–13 mm long, 4–6.5 mm wide
Petals 9.5–13 mm long, 4–6 mm wide
Lip 9–11.5 mm long, 8–9.5 mm wide
Column 5.5–6.5 mm long

Pollardia venosa
FIGURES 7-14, 7-15, PLATES 82, 83

Pollardia venosa (Lindley) Withner & Harding, *comb. nov.* Basionym: *Epidendrum venosum* Lindley. 1831. *Gen. Sp. Orch. Pl.* 99.

SYNONYMS

Epidendrum ensicaulon A. Richard & Galeotti. 1845. *Ann. Sci. Nat.*, ser.3, 3: 19.
Epidendrum wendlandianum Kraenzlin. 1893. *Gard. Chron.*, ser.3, 13: 58.
Encyclia wendlandiana (Kraenzlin) Schlechter. 1918. *Beih. Bot. Centralbl.* 36 (2): 474.
Encyclia venosa (Lindley) Schlechter. 1918. *Beih. Bot. Centralbl.* 36 (2): 474.
Prosthechea venosa (Lindley) W. E. Higgins. 1997. *Phytologia* 82 (5): 381.

DERIVATION OF NAME

Latin *venosus*, "branched veins," referring to the lip.

DESCRIPTION

Pseudobulbs spaced 1–4 cm apart on the rhizome, ovoid-ellipsoid to spindle-shaped. Leaves two or three. Inflorescence with two to five resupinate flowers. Sepals and petals pale green more or less striped with red-brown, lip white lateral lobes with violet lines. Sepals and petals lanceolate, pointed. Lip three-lobed, adnate to column for about a third of the length, lateral lobes triangular to oblong, obtuse to bluntly acute, separated from midlobe by deep sinus on each side, midlobe broadly triangular, apex obtuse or broadly acute. Callus fleshy, pubescent, becoming three to five fleshy veins that run the length of the midlobe. Column blue-violet, teeth yellow. Capsule three-winged.

HABITAT AND DISTRIBUTION
México. Found at 30–2200 meters, in pine-oak forest, occasionally on rocks.

FLOWERING TIME
April to September.

COMMENT
Photographs of several color forms were submitted for this book. All the plants appear to be *Pollardia venosa*, but they do show that the species is variable. Two of these forms are shown in Plates 82 and 83.

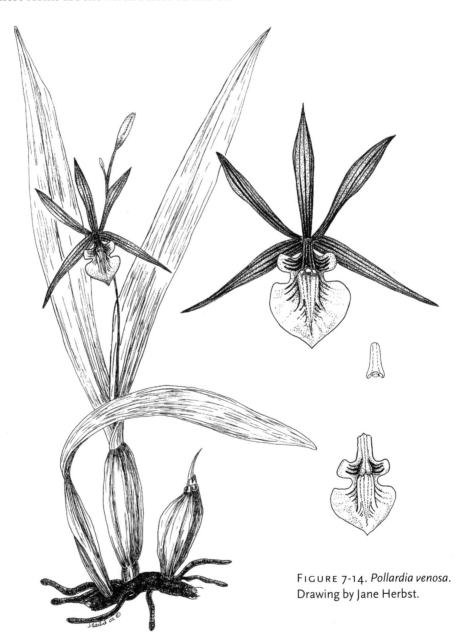

FIGURE 7-14. *Pollardia venosa*. Drawing by Jane Herbst.

MEASUREMENTS

Pseudobulbs 3–7.5 cm long, 0.6–2.2 cm wide
Leaves 10–22 cm long, 0.6–1.9 cm wide
Inflorescence 6–12 cm long
Sepals 19–27 mm long, 2–4 mm wide
Petals 16–25 mm long, 1–4 mm wide
Lip 14–19.5 mm long, 7–14 mm wide
Column 6–8 mm long

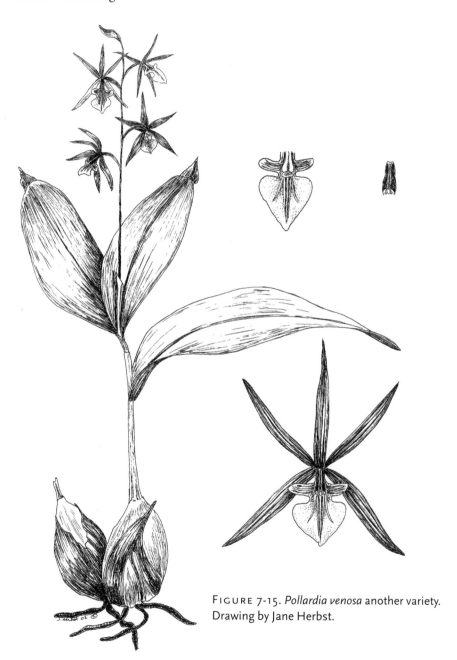

FIGURE 7-15. *Pollardia venosa* another variety. Drawing by Jane Herbst.

Encyclia pastoris

Encyclia pastoris (Llave & Lexarza) Schlechter. 1918. *Beih. Bot. Centralbl.* 36 (2): 473. Basionym: *Epidendrum pastoris* Llave & Lexarza. 1825. *Novorum Vegetabilium Descriptiones* Fasc. 2: 23.

SYNONYM

Prosthechea pastoris (Llave & Lexarza) Espejo & Lopez-Ferrari. 2000. *Acta Botanica Mexicana* 51: 61.

COMMENT

A name you will see in the literature is *Encyclia pastoris*. It has been listed as a synonym of *Pollardia venosa* and of other species by various authors. We have chosen to leave it out of the above lists because we do not know to which species it belongs. We include the literature citations here.

According to Dressler (1974), the type description is vague and it could refer to *Pollardia linkiana* or *P. pterocarpa*. Östlund considered this name to be the correct name for *P. venosa*.

CHAPTER 8

The Genus *Prosthechea*

The proportions of the lip structures of *Prosthechea* flowers seem to be markedly different from those of the other genera of the *Epidendrum* complex. There is a significantly longer claw at the base of the lip, the lateral lobes of the lip are significantly less developed or absent. Some have a callus that is scarcely developed and is mostly just a few thickened veins, others have a callus that is very thick. Most of the species are comparatively rare and are not well known or recognized, with only sixteen species described at present in the genus as accepted here in a narrow sense. Many of these species are also distinguished by their glaucous or silvery appearance. The genus is found in both Central and South America. *Prosthechea* flowers are mostly resupinate and seem to lack the crystal bundles that are indicative of the cells of *Anacheilium* species.

Lindley included *Prosthechea* under *Epidendrum* section *Encyclia* in his key to the subsections of *Epidendrum*, not considering it to be a close relative of *Anacheilium* or *Hormidium;* however in its current use today, it seems to take in species that are considered close relatives to these genera.

Other complications to the genus name date back to the first use of this generic epithet by Knowles and Westcott in 1838 (*The Floral Cabinet* 2: 111), when they described *Prosthechea glauca* "nearly related to the genus *Epidendrum*, but differs in the structure of the labellum and column... the generic name... in reference to the appendage on the back of the column." The plant had come from México, was covered "with a glaucous hue," and the clinandrium was "roundish, entire, apiculate at the base, and just below is a roundish fleshy appendage streaked with purple." The lip was "linear-lanceolate and pressed to the column, fleshy, and more especially at the apex, which is divided into three unequal lobes." The purplish anther cap covered four somewhat pear-shaped pollinia that were situated at the back (tip) of the column and had four caudicles. This was distinctive enough, and these characteristics, particularly the appendage, did warrant a separation from *Epidendrum*.

In the same volume of *The Floral Cabinet* (p. 167), in which their *Prosthechea* appeared, Knowles and Westcott withdrew their new generic name because it was, in their opinion, too similar to another based on the same Greek origin. They then substituted *Epithecia* as a preferred, less confusing epithet for their Mexican genus, the epithet derived from a different Greek word with approximately the same meaning. The name *Prosthechea*, however, remains in the orchid literature and is retained because of the rule of page priority, although it may be a little clouded.

Higgins (1997), from his work on a cladistic study of *Encyclia* based in part on molecular studies, has transferred more than ninety species into *Prosthechea*, apparently without noting some key morphological characteristics in comparison to the numerical cladistic information.

Even the type specimen *Prosthechea glauca* has problems with its nomenclature. The name *glauca* was invalid when transferred to the genus *Epidendrum* since Swartz had already used *Epidendrum glaucum* in 1788 in his *Prodromus* for a species from Cuba, and the Cuban species was different from the Mexican entity described by Knowles and Westcott. Ames and his colleagues, recognizing the problem, renamed the Knowles and Westcott species *Epidendrum glaucovirens* in 1934. The Caribbean species of Swartz is now called *Dichaea glauca*. There is no problem with using the species name *glauca* in the genus *Prosthechea*.

Withner (2001) listed only five species for *Prosthechea*: *P. christyana*, *P. fortunae*, *P. glauca*, *P. serpentilingua*, and *P. squamata*. The paper also suggested that

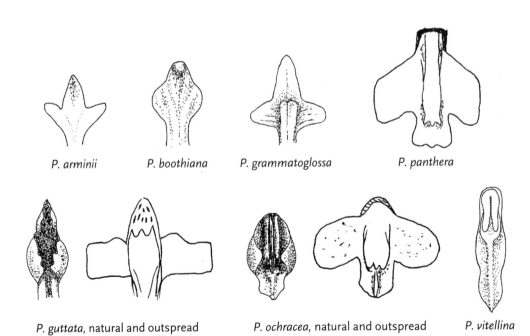

FIGURE 8-1. *Prosthechea* lips.

most of the ninety-two species transferred to this genus by Higgins (1997) be returned to their original or more appropriate genera.

We now include other species in this group, but admit that the fit is a bit strained, and we also find that these species fit into three or four groups within this genus. In the future, it may be that these groups will become separate genera.

Species of *Prosthechea*

Prosthechea arminii, 254
Prosthechea bicamerata, 255
Prosthechea boothiana, 256
Prosthechea brachiata, 259
Prosthechea christyana, 260
Prosthechea cretacea, 261
Prosthechea fortunae, 262
Prosthechea glauca, 263
Prosthechea grammatoglossa, 265
Prosthechea guttata, 267
Prosthechea magnispatha, 269
Prosthechea ochracea, 270
Prosthechea ortizii, 272
Prosthechea panthera, 273
Prosthechea serpentilingua, 274
Prosthechea vitellina, 275

Key to Species of *Prosthechea*

1a. Lateral lip lobes larger than medial lobe . . . . go to 9
1b. Lateral lip lobes smaller than medial lobe, or lip without lateral lobes . . . . . . go to 2
2a. Midtooth of column shorter than lateral teeth . . . . go to 3
2b. Midtooth of column not shorter than lateral teeth, flower color not orange . . . . go to 5
3a. Lip one-lobed . . . . *Prosthechea vitellina*
3b. Lip three-lobed . . . . go to 4
4a. Lip with no callus, flower brown or yellow-brown . . . . *Prosthechea arminii*
4b. Lip with callus of two keels, flower cream yellow to green . . . . *Prosthechea grammatoglossa*
5a. No callus on lip . . . . go to 6
5b. Callus distinct . . . . go to 7
6a. Lip unlobed . . . . *Prosthechea serpentilingua*
6b. Lip three-lobed . . . . *Prosthechea christyana*
7a. Midlobe of lip not folded under . . . . *Prosthechea fortunae*
7b. Midlobe of lip at right angle or folded under column . . . . go to 8
8a. Callus triangular in shape with platelike keel . . . . *Prosthechea ortizii*
8b. Callus merely fleshy . . . . *Prosthechea glauca*
9a. Midtooth of column surpassing lateral teeth . . . . go to 10
9b. Midtooth of column not surpassing lateral teeth . . . . go to 11
10a. Petals 8–10 mm long . . . . *Prosthechea boothiana*
10b. Petals 18–22 mm long . . . . *Prosthechea magnispatha*

11a. Column teeth digitate, fimbriate, not rounded but equal length flowers cupped... .... go to 12
11b. Teeth subequal, rounded, fleshy, callus thick with fleshy parallel furrows (seed capsule not winged, triangular) .... go to 14
12a. Midlobe lip wider at apex than at base .... *Prosthechea panthera*
12b. Midlobe of lip not wider at apex than base .... go to 13
13a. Sepals with spiny warty outer/reverse .... *Prosthechea guttata*
13b. Sepals smooth on reverse .... *Prosthechea ochracea*
14a. Callus of lip running into three fleshy keels .... *Prosthechea cretacea*
14b. Callus not running into three keels .... go to 15
15a. Sepals and petals widest near base .... *Prosthechea bicamerata*
15b. Sepals and petals widest near apex .... *Prosthechea brachiata*

Prosthechea arminii

Prosthechea arminii (Reichenbach f.) Withner & Harding, *comb. nov.* Basionym: *Epidendrum arminii* Reichenbach f. 1855. *Bonplandia* 3: 67.

SYNONYM
Encyclia arminii (Reichenbach f.) Carnevali & I. Ramirez. 1986. *Ernestia* 36: 9.

DESCRIPTION
A very small epiphyte or a lithophyte. Pseudobulbs spindle-shaped. Leaf one. Inflorescence simple, with basal sheath. Flowers six non-resupinate. Tepals brown, lip yellow with dark purple lines. Lip has fine white hairs, though there is no callus. The drawing shows the lateral column teeth extending more distally than the medial tooth.

HABITAT AND DISTRIBUTION
Colombia and Venezuela. Found at 800–1250 meters in elevation.

MEASUREMENTS
Pseudobulbs 1 cm long, 0.2 cm wide
Leaves 1.5 cm long
Inflorescence 1.5 cm long
Spathe 0.1 cm long
Sepals 8 mm long, 1 mm wide
Petals 6 mm long, 1 mm wide
Lip 6 mm long, 5 mm wide
Column 3 mm long

Prosthechea bicamerata
PLATE 84

Prosthechea bicamerata (Reichenbach f.) W. E. Higgins. 1997. *Phytologia* 82 (5): 381. Basionym: *Epidendrum bicameratum* Reichenbach f. 1871. *Gardener's Chronicle Agricultural Gazette,* ser.1, 28: 1194.

SYNONYM
Encyclia bicamerata (Reichenbach f.) Dressler & G. E. Pollard. 1971. *Phytologia* 21 (7): 436.

DERIVATION OF NAME
Latin *bi,* "two," and *camera,* "chambered."

DESCRIPTION
Pseudobulbs clustered, ovoid. Leaves two or three. Flowers eleven to twenty-six, resupinate. Sepals and petals dark yellowish brown or chestnut brown, lip white becoming yellow with age, with a few purple dots at apex of callus. Lip basally adnate to column, three-lobed, lateral lobes quadrate (square), clasped over column, midlobe oblong or kidney-shaped. Callus oblong, massive, sulcate. Column teeth flat and thin, subequal. Capsule bluntly triangular in cross section but not winged.

HABITAT AND DISTRIBUTION
México. Found at 2100–2600 meters in elevation, in wet mixed forest or cloud forests.

FLOWERING TIME
September to April.

MEASUREMENTS
Pseudobulbs 2.5–6.5 cm long, 2.5–3.2 cm wide
Leaves 12–24.5 cm long, 1.5–3.4 cm wide
Inflorescence 15–39 cm long
Spathe 3–3.5 cm long
Sepals 11–13.5 mm long, 5–6.5 mm wide
Petals 10–12 mm long, 4.2–6 mm wide
Lip 8–10 mm long, 5.5–6.5 mm wide
Column 5–6 mm long

Prosthechea boothiana
FIGURE 8-2, PLATE 85

Prosthechea boothiana (Lindley) W. E. Higgins. 1997. *Phytologia* 82 (5): 38. Basionym: *Epidendrum boothianum* Lindley. 1838. *Edward's Botanical Register* 24: 5. Type: Cuba, Havana, 1835, *Sutton s.n.* (holotype: K).

SYNONYMS
Epidendrum bidentatum Lindley. 1831. *Gen. Sp. Orch. Pl.* 98. not *E. bidentatum* König.
Epidendrum favoris Reichenbach f. 1874. *Gard. Chron.*, n.s. 2: 98.
Diacrium bidentatum (Lindley) Hemsley. 1883. *Biol. Centr-Amer. Bot.* 3: 221.
Epidendrum erythronioides Small. 1903. *Fl. Southeastern U.S.* 328. 1329. Type: USA, Florida, Key Largo, *Curtiss s.n.* (holotype: NY).
Epicladium boothianum (Lindley) Small. 1913. *Fl. Miami* 56.
Hormidium boothianum (Lindley) Brieger. 1960. *Publicacao Cientifica Universidade de São Paulo, Institut de Genetica*, 1: 21.
Encyclia boothiana (Lindley) Dressler. 1961. *Brittonia* 13 (3): 264.
Encyclia boothiana subsp. *boothiana*. 1961. *Brittonia* 13 (3): 264.
Epidendrum boothianum var. *erythronioides* (Small) Luer. 1971. *Florida Orchidist* 14: 29.
Encyclia boothiana subsp. *favoris* (Reichenbach f.) Dressler & Pollard. 1971. *Phytologia* 21: 436.
Encyclia bidentata subsp. *erythronioides* (Small) Hágsater. 1993. *Orquídea (México)* 13 (1–2): 216.
Encyclia bidentata (Lindley) Hágsater & Soto Arenas. 1993. *Orquídea (México)* 13 (1–2): 215–218.
Encyclia boothiana subsp. *erythronioides* (Small) Hágsater ex Christenson. 1996. *Lindleyana* 11 (1): 15.
Prosthechea boothiana var. *erythronioides* (Small) W. E. Higgins. 1999. *North American Native Orchid Journal* 5 (1): 18.

COMMON NAME
Dollar orchid, because of the round, extremely flattened pseudobulbs that resemble an old-fashioned silver dollar.

DESCRIPTION
Pseudobulbs clustered, extremely flattened. Leaves two. Inflorescence unbranched with one to five resupinate flowers. Sepals and petals yellow heavily spotted (barred transversely) with chestnut brown, lip greenish yellow. Lip adnate to col-

umn basally, blade rhombic (sides of equal length) or weakly three-lobed, lateral lobes rounded or blunt and turned downward, apex of midlobe rounded. Callus fleshy, with apex thickened to fleshy, oblong. Column bowed, midtooth surpassing lateral teeth. Capsule three-winged.

HABITAT AND DISTRIBUTION

Belize, Florida, Honduras, México, and the West Indies. Found in dry scrub forest and tropical deciduous forest, at 150 meters in elevation (excluding subsp. *favoris*, which is found at higher elevations).

FIGURE 8-2. *Prosthechea boothianum*. Drawing by Jane Herbst.

FLOWERING TIME
At different times depending on the subspecies.

CULTURE
Greg Allikas grows this plant well in Florida. We have had no luck with it, so we defer to his culture report.

> We (other Floridian growers and I) have noticed that there seem to be two forms of this species. I have not compared the two side by side, but have had both in cultivation at one time or another. The Florida form has many small flowers whereas the Belize form has fewer larger flowers. The size difference is somewhat drastic, yet the flowers and the plants appear similar in both types. We grow *Prosthechea boothiana* mounted on cork. It receives bright light hung low in the *Vanda* area, which is covered with 43 percent shade cloth. Our particular plant seems to be somewhat deciduous and is of the large flowered form. We do nothing special but that section of the shadehouse receives water every day. We fertilize once a week during warm months and every other week in winter using 20–20–20 in a weak dilution. Inflorescences are produced as the new growth matures.

COMMENTS
Prosthechea boothiana var. *erythronioides* (Small) Higgins, the only subspecies to occur in Florida, has three anthers. Plants with three anthers are found rarely elsewhere, in mixed populations with normal plants.

In 2001 Salazar and Soto Arenas (*Lindleyana* 16 [3]: 149) raised *Prosthechea favoris* to the species level based on geography and morphology. It is recognized by the same traits, only larger with wider sepal and petals; it is found in an area geographically separate from that of *P. boothiana*, in México at 800–1500 meters in elevation in wet mixed hardwood forest, and it blooms in September and April to May.

MEASUREMENTS
Pseudobulbs 2–3.5 cm long (subsp. *favoris* is larger), 1.8–2.5 cm wide
Leaves 5–13 cm long, 1–2 cm wide
Inflorescence 5–10 cm long
Spathe 2–4.5 cm long
Sepals 9–11 mm long, 3.5–4 mm wide
Petals 8–10 mm long, 2–4 mm wide
Lip 7–8 mm long, 6–7 mm wide
Column 5–5.5 mm long

Prosthechea brachiata
PLATE 86

Prosthechea brachiata (A. Richard & Galeotti) W. E. Higgins. 1997. *Phytologia* 82 (5): 381. Basionym: *Epidendrum brachiatum* A. Richard & Galeotti. 1945. *Annales des Sciences Naturelles; Botanique*, sér. 3, 3: 20.

SYNONYM
Encyclia brachiata (A. Richard & Galeotti) Dressler & G. E. Pollard. 1971. *Phytologia* 21 (7): 436.

DERIVATION OF NAME
Latin *brachiatus,* "branching alternately in opposite directions," referring to the inflorescence.

DESCRIPTION
Pseudobulbs loosely clustered, ovoid, strongly flattened. Leaf one. Inflorescence simple or branched, with thirteen to twenty-four flowers. Sepals and petals greenish yellow on the reverse, the inner face yellow blotched with red-brown, lip yellow with red-brown dots on midlobe. Lip basally adnate to column, three-lobed, lateral lobes oblong, narrowly rounded, midlobe basally wedge-shaped, oval. Callus fleshy, oblong, sulcate. Column teeth fleshy, subequal or midtooth longer. Capsule slightly triangular in cross section.

HABITAT AND DISTRIBUTION
México. Found at 1700–2200 meters in elevation, in rather wet pine-oak forests.

FLOWERING TIME
September to April.

MEASUREMENTS
Pseudobulbs 4 cm long, 1.5 cm wide
Leaves 15–20 cm long, 3.5–3.8 cm wide
Inflorescence 15–25 cm long
Sepals 6.5–7.5 mm long, 4 mm wide
Petals 5.5–7 mm long, 3.5 mm wide
Lip 6.5–8 mm long, 2.3–2.5 mm wide
Column 4–4.5 mm long

Prosthechea christyana

Prosthechea christyana (Reichenbach f.) Garay and Withner, *comb. nov.* Basionym: *Epidendrum christyanum* Reichenbach f. 1884. *Gard. Chron.*, n.s., 22: 38.

SYNONYMS

Epidendrum squamatum Barbosa Rodrigues. 1882. *Gen. Sp. Orch. Nov.* 2: 134. Type: Brazil. Lectotype: Rodrigues' original drawing (see S. Sprunger et al., eds. 1996. J. Barbosa Rodrigues' *Iconographie des Orchidées du Brasil* 1: 273).
Epidendrum megahybos Schlechter. 1929. *Repertorium Specierum Novarum Regni Vegetabilis, Beihefte* 57 (8): 75.
Encyclia squamata Porto & Brade. 1935. *Rodriguesia* 1 (2): 29.
Epidendrum hoehnei Hawkes. 1957. *Orquídea* 18 (5): 176.
Encyclia hoehnei (Hawkes) Pabst. 1975. *Bradea* 2: 21.
Encyclia megahybos (Schlechter) Ortiz. 1991. *Orquideologia* 18 (1): 99.
Prosthechea squamata (Porto & Brade) W. E. Higgins. 1997. *Phytologia* 82 (5): 370.
Prosthechea megahybos (Schlechter) Dodson & Hágsater. 1999. *Monographs in Systematic Botany* 75: 956.

DERIVATION OF NAME

Latin *squamatum*, "scaly," referring to the lip.

DESCRIPTION

The major feature of this species is the three-lobed lip with the dorsally scaly base (Rodrigues said *squamose* meaning "scaly"). Unfortunately, an earlier orchid species was given the name *Epidendrum squamatum*, so the name is not available for this species. This has resulted in a variety of name changes over the years as you can see under the synonyms. The earlier name, *E. squamatum*, was used by J. L. M. Poiret (1810) when he transferred *Ophrys squamatum* G. Forester, a species from New Caledonia, to *Epidendrum*. This species is now known as *Dipodium squamatum*.

Another feature of the lip structure is the elongated claw or lip base, and the lack of a forcipate callus. Reichenbach called the species "a great botanical curiosity" because of its unusual features. It would appear to have not been collected a second time since its original collection and naming by Barbosa Rodrigues, resulting in further synonyms. There are many species of orchids for which this is presently the case—a real concern for those interested in plant identification and the conservation of native habitats, as well as a job for the herbaria where the type specimens are conserved for the future, in that these synonyms and basionyms need to be sorted out.

Fortunately, we have, for this species, the original drawing of the plant and flower by Rodrigues. We also have Schlechter's drawing of *Prosthechea megahybos*, now considered synonymous. It is interesting to note that the inturned edges of the column hold the lip tightly in place and form two pointed extensions on either side of the anther. There is also a flaplike structure at the tip of the column. The drooping flower stalk produces only a few flowers, and it does not extend beyond the two leaves. The pseudobulbs are narrowly pyriform with fine ridges. The fruit is a three-sided capsule.

HABITAT AND DISTRIBUTION
Bolivia, Brazil, Ecuador, and Peru.

COMMENT
The type specimen was brought by T. Christy from Bolivia. It is conserved in the Reichenbach Herbarium in Vienna (#31708 flower, #118 drawing). The lectotype of *Prosthechea megahybos* is Schlechter's original drawing published by R. Mansfeld in *Repertorium Specierum Novarum Regni Vegetabilis, Beihefte* 57: t. 89, no. 347, 1929. It was collected in Ecuador, Chimborazo Province, at Pallatanga by Sodiro (isotype possibly in Brussels, BR).

Prosthechea cretacea

Prosthechea cretacea (Dressler & Pollard) W. E. Higgins. 1997. *Phytologia* 82 (5): 381. Basionym: *Encyclia cretacea* Dressler & Pollard. 1971. *Phytologia* 21 (7): 438. Type: México, Oaxaca, 2.8 kilometers from Tuxtepec highway at Kilometer 21, east of highway on logging road, on oaks in pine-oak forest, at 2475 meters, 16 January 1971, *G. E. Pollard s.n.* (holotype: US).

DERIVATION OF NAME
Latin *cretaceous*, "chalky," referring to the powdery coating of the plants.

DESCRIPTION
Pseudobulbs clustered, ovoid, slightly flattened, younger pseudobulbs with whitish sheaths. Leaves two or three, light green with whitish powdery coating. Flowers ten to twenty-four. Sepals and petals dull green to olive green, becoming yellow, lip white, becoming yellow. Sepals fleshy. Lip basally adnate to column, three-

lobed, lateral lobes linear lanceolate, narrowly obtuse, midlobe clawed with wide isthmus, triangular cordate. Callus fleshy, running into three fleshy keels, midkeel nearly reaching the apex of lip. Column teeth short, obtuse, subequal. Capsule faintly triangular in cross section but rounded, not keeled or winged. Autogamous (self-pollinated).

HABITAT AND DISTRIBUTION
México. Found at 2400–2500 meters in elevation, in pine-oak forest often bathed in clouds. Grows on underside of branches; plants hang vertically downward toward ground.

FLOWERING TIME
November to March.

CULTURE
Culture requirements are unknown to us.

MEASUREMENTS
Pseudobulbs 7–9 cm long, 4–4.8 cm wide
Leaves 16–32.5 cm long, 3–4.3 cm wide
Inflorescence 18–36 cm long
Spathe 3 cm long
Sepals 12–13 mm long, 4–5 mm wide
Petals 11–12 mm long, 3 mm wide
Lip 12 mm long, 6–6.5 mm wide
Column 5.5–6 mm long

Prosthechea fortunae

Prosthechea fortunae (Dressler) W. E. Higgins. 1997. *Phytologia* 82 (5): 277. Basionym: *Encyclia fortunae* Dressler. 1980. *Orquídea (México)* 7 (4): 357. Type: Panama, Chiriqui Province, near Cam Hornito, Fortuna Dam Site, *Dressler 5520* (holotype: US; isotype: PMA).

DERIVATION OF NAME
After Fortuna Dam, in Panama.

DESCRIPTION
Entire plant is whitish-glaucous, pendent. Pseudobulbs clustered, ovate to suborbicular, strongly flattened. Leaves one or two. Flowers six to eight. Sepals and petals pale green, lip cream. Lip three-lobed, lateral lobes orbicular, erect, midlobe

pale green. Callus fleshy, faintly three-toothed between lateral lobes. Column with three apical teeth.

HABITAT AND DISTRIBUTION

Panama.

CULTURE

Dressler said, "In the very unlikely event that anyone ever wished to cultivate *Encyclia fortunae*, cool, moist conditions are indicated by its habitat." The entire plant was glaucous of a waxy white color, pendent, and grew with clustered pseudobulbs, having one or two leaves and six to eight flowers. The callus was fleshy and faintly three-toothed between the lateral lobes. The midlobe of the lip does not fold under as does that of *Prosthechea glauca*.

COMMENT

This species is related to *Prosthechea glauca* but distinguished by its smaller flowers and midlobe of lip that does not fold under. *Pollardia campylostalix* also occurs in the same geographic area but has much larger flowers and a longer column and lip.

Little is known of this species, and as far as we can find, it has not been collected or grown by hobbyists. It was described originally as an *Encyclia* with greenish flowers and a creamy lip. The sepals are recorded as 6 mm long. According to our research, it belongs in this genus because of its lip configuration. Until we know more about the species there is little more to say.

MEASUREMENTS

Pseudobulbs 2 cm long, 1.2–1.6 cm wide
Leaves 9–10.5 cm long, 1.4–2.3 cm wide
Inflorescence 7–10 cm long
Spathe 4–7 mm long
Sepals 6 mm long, 2.3–2.8 mm wide
Petals 5.1 mm long, 2 mm wide
Lip 5.3 mm long, 2 mm wide
Column 4 mm long

Prosthechea glauca

PLATES 87, 88

Prosthechea glauca Knowles & Westcott. 1838. *Floricultural Cabinet and Florist's Magazine* 2: 111. Type: México, without proper locality. Holotype: Imported and cultivated by *G. Barker s.n.* (K-Lindley).

SYNONYMS

Epithecia glauca Knowles & Westcott. 1839. *Floral Cabinet* 2: 167, t. 87.

Epidendrum glaucum (Knowles and Westcott) Lindley. 1840. *Edward's Botanical Register* 26, misc. 29, not Swartz, 1788.

Epidendrum limbatum Lindley. 1843. *Edward's Botanical Register* 29, misc. 69. Type: Guatemala, without precise locality. Holotype: *Skinner s.n.* (K-Lindley).

Amblostoma tridactylum var. *mexicanum* Kraenzlin. 1920. *Vidensk. Medd. Dansk Naturh. Foren.* 71: 177. Holotype: México, without precise locality, *Liebman s.n.*, conserved in Copenhagen (C).

Epidendrum glaucovirens Ames, Hubbard & Schweinfurth. 1935. *Botanical Museum Leaflets* 3 (5): 70.

Encyclia limbata (Lindley) Dressler. 1961. *Brittonia* 13 (3): 265.

Encyclia glauca (Knowles & Westcott) Dressler & G. E. Pollard. 1971. *Phytologia* 21 (7): 437.

DERIVATION OF NAME

Latin *glaucous*, "covered with a waxy bloom," referring to the plant's surface.

DESCRIPTION

Pseudobulbs roundish oval, compressed (flattened). Leaf one. Inflorescence a pedunculate raceme with the flowers spreading in the manner of an *Epidendrum*. Sepals ovate, acute, purplish tipped with greenish-yellow. Petals about one-third the size of the sepals, lanceolate, acute, similar in color to sepals. Lip linear-lanceolate, three-lobed, parallel and pressed to column, fleshy, the lobes unequal, central lobe broad, lateral lobes longer and narrower with fleshy angular appendage at the back. Column curved with an angular hollow at base. Clinandrium roundish, entire, apiculate at base with roundish fleshy appendage immediately below, streaked with purple. Plants are entirely covered with a glaucous hue.

HABITAT AND DISTRIBUTION

México, Guatemala. Found at 900–1000 meters in elevation, in moderately wet oak or mixed forests.

FLOWERING TIME

April to August.

COMMENT

There is no type specimen for this species; however, there is a detailed color plate from the *Floral Cabinet*, which we have included as Plate 87.

MEASUREMENTS

Pseudobulbs 2.5–6 cm long, 2–3.5 cm wide
Leaves 9–18 cm long, 2.2–4 cm wide
Inflorescence 20–35 cm long
Sepals 6.5–7.5 mm long, 2.8–3.5 mm wide
Petals 5–5.8 mm long, 1.8–2 mm wide
Lip 6.5 mm long, 2–2.5 mm wide
Column 3.5–4 mm long

Prosthechea grammatoglossa
FIGURE 8-3, PLATE 89

Prosthechea grammatoglossa (Reichenbach f.) W. E. Higgins. 1997. *Phytologia* 82 (5): 381. Basionym: *Epidendrum grammatoglossum* Reichenbach f. 1849. *Linnaea* 22: 837.

SYNONYMS

Epidendrum quadridentatum Lehmann & Kraenzlin. 1899. *Engl. Bot. Jahrb. Syst.* 26: 459.
Encyclia grammatoglossa (Reichenbach f.) Dressler. 1961. *Brittonia* 13: 265.
Anacheilium grammatoglossum (Reichenbach f.) Pabst, Moutinho & A. V. Pinto. 1981. *Bradea* 3: 23.

DERIVATION OF NAME

Latin *grammatus*, "striped with raised lines," and *glossum*, "tongue," referring to the lip.

DESCRIPTION

An epiphyte or terrestrial on mossy rocks. Rhizome creeping. Pseudobulbs variable, surrounded by sheath at base. Leaves two. Flowers four to eight, cream yellow to green. Petals lower margins involute. Lip bluish green, lateral lobes with three or four purple veins, column greenish yellow. Lip adnate for half the length of the column. Lip trilobed, lateral lobes erect (incurved) folding over column, central lobe triangular, disc with two oblong calli, slightly concave with four unequal magenta stripes on disc. Column stout, with two slender triangular incurved stigmatic wings. Anthers orange. Lateral teeth of column longer than midtooth.

HABITAT AND DISTRIBUTION
Bolivia, Colombia, Ecuador, Peru, and Venezuela. Found at 750–820 meters in elevation in wet tropical cloud forest.

FLOWERING TIME
December to May.

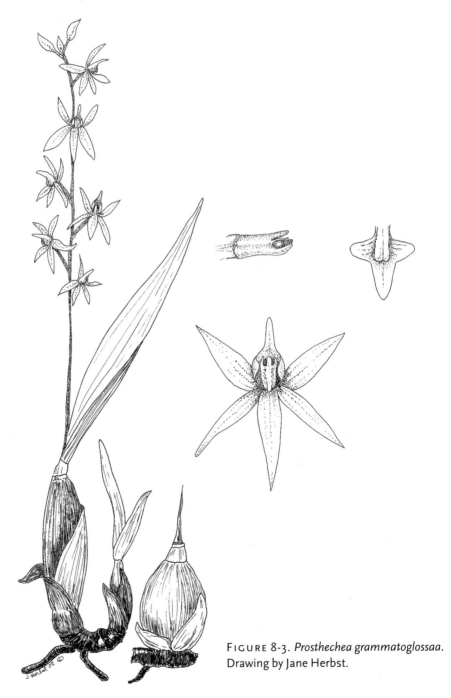

FIGURE 8-3. *Prosthechea grammatoglossaa*. Drawing by Jane Herbst.

CULTURE
Culture for this plant would be wet, humid conditions and an intermediate growing environment.

MEASUREMENTS
Pseudobulbs 2.5–8 cm long, 1 cm wide
Leaves 17.6 cm long, 0.6–1.2 cm wide
Inflorescence 10–15 cm long
Sepals 9 –10 mm long, 2.3 mm wide
Petals 7 mm long, 1.5 mm wide
Lip 5.2 mm long, 5 mm wide
Column 3–4 mm long, 2 mm wide

Prosthechea guttata
FIGURE 8-4, PLATE 90

Prosthechea guttata (Schlechter) Christenson. 2003. *Richardiana* 3(3): 116. Basionym: *Encyclia guttata* Schlechter. 1918. *Beih. Bot. Centralbl.* 36 (2): 472. Type: México. Oaxaca, Cerro San Felipe, *H. Galeotti 5029* (holotype: P; isotype: W).

SYNONYMS
Epidendrum guttatum A. Richard & H. Galeotti.1845. *Ann Sci. Nat.*, ser. 3, 3: 20, *nom illeg.*, non Linnaeus.
Epidendrum maculosum Ames, F. T. Hubbard & C. Schweinfurth. 1935. *Botanical Museum Leaflets* 3: 72.
Encyclia maculosa (Ames, F. T. Hubbard & C. Schweinfurth) Hoehne. 1952. *Arquivos de botanica do estado de São Paulo*, 2: 1552.
Hormidium guttatum (A. Richard & H. Galeotti) Brieger. 1977. Schlechter's *Die Orchideen*, ed. 3, p. 573.
Prosthechea maculosa (Ames, F. T. Hubbard & Schweinfurth.) W. E. Higgins. 1997. *Phytologia* 82 (5): 381, *nom. illeg.*

DERIVATION OF NAME
Latin *guttatus*, "spotted."

DESCRIPTION
Pseudobulbs loosely clustered, up to 2 cm apart, conic-ovoid to spindle-shaped. Leaves two or three. Inflorescence unbranched, with three to twenty non-resupinate flowers. Sepals and petals orange or brownish orange, dotted with red-brown, lip white with purple-maroon dots. Sepals echinate (warty) on reverse

side. Lip adnate to basal quarter of column, three-lobed, lateral lobes clasping column, oblong, midlobe triangular-ovate, obtuse. Callus thick, three-toothed with each tooth fimbriate distally, midlobe of lip with three crenulate veins, midvein running to apex. Column three-toothed. Capsule three-winged.

HABITAT AND DISTRIBUTION

México. Found at 1400–2000 meters in elevation, epiphytic on oak, dry to fairly wet.

FIGURE 8-4. *Prosthechea guttata*. Drawing by Jane Herbst.

FLOWERING TIME

March to July.

COMMENT

The species is similar to *Prosthechea ochracea*, but all three lobes of the lip are usually narrower, the sepals and petals are dotted with brown, the sepals have warts on the outside (reverse), and the leaves tend to be wider.

MEASUREMENTS

Pseudobulbs 4–12 cm long, 1–2.7 cm wide
Leaves 6–28 cm long, 0.9–2 cm wide
Inflorescence 6–20 cm long
Spathe 3 cm long
Sepals 4.8–7 mm long, 2–5 mm wide
Petals 5–6.5 mm long, 2–3 mm wide
Lip 4.5–6.5 mm long, 2–3.2 mm wide
Column 3.5–5 mm wide

Prosthechea magnispatha
PLATE 91

Prosthechea magnispatha (Ames, F. T. Hubbard & C. Schweinfurth) W. E. Higgins. 1997. *Phytologia* 82 (5): 381. Basionym: *Epidendrum magnispathum* Ames, F. T. Hubbard & C. Schweinfurth. 1934. *Botanical Museum Leaflets* 3: 10. Type: México, State of Guerrero, January 1933, flowered at Cuernavaca, 23 November 1933. leg. O. Nagel, *Erik M. Östlund 2043* (Type in Herb. Östlund; flowers from type plant in AMES).

SYNONYM

Encyclia magnispatha (Ames, F. T. Hubbard & C. Schweinfurth) Dressler. 1961. *Brittonia* 13: 265.

DERIVATION OF NAME

Latin magni, "big," and spathum, "spathe," referring to the big spathe enclosing the base of the inflorescence.

DESCRIPTION

Pseudobulbs clustered, strongly flattened. Leaves two or three. Inflorescence unbranched with four to seven flowers. Sepals and petals yellowish green heavily spotted with chocolate brown. Lip cream or yellow. Lip adnate to basal third of the column, blade rhombic (sides of equal length) or weakly three-lobed, lateral lobes rounded or blunt and turned downward (margins not turned inward), apex rounded. Callus fleshy, oblong. Column bowed, midtooth surpassing lateral teeth. Capsule three-winged.

HABITAT AND DISTRIBUTION
México. Found at 800–1800 meters in elevation, in warm to hot wet montane oak and mixed forests.

FLOWERING TIME
Fall and winter.

COMMENT
The species is similar to *Prosthechea boothiana* except is larger, has blunter perianth segments, differently shaped petals, and the lip midlobe is not emarginate.

MEASUREMENTS
Pseudobulbs 6–9.5 cm long, 4–7 cm wide
Leaves 10–26 cm long, 3–5 cm wide
Inflorescence 12–30 cm long
Spathe 7–15 cm long
Sepals 19–24 mm long, 5.5–7 mm wide
Petals 18–22 mm long, 5–7 mm wide
Lip 14–15 mm long, eight–9 mm wide
Column 11–12 mm wide

Prosthechea ochracea
FIGURE 8-5, PLATE 92

Prosthechea ochracea (Lindley) W. E. Higgins. 1997. *Phytologia* 82 (5): 381. Basionym: *Epidendrum ochraceum* Lindley. 1838. *Edward's Botanical Register* 24, misc.15, t. 26.

SYNONYMS
Epidendrum triste A. Richard & H. Galeotti. 1845. *Ann. Sci. Nat.*, ser. 3, 3: 20.
Epidendrum parviflorum Sessé & Mociño. 1894. *Flora Mexicana*, ed. 2: 206.
Encyclia ochracea (Lindley) Dressler. 1961. *Brittonia* 13 (3): 265.

DERIVATION OF NAME
Latin *ochraceus*, "yellowish," referring to the flower color.

DESCRIPTION

Pseudobulbs loosely clustered, ovoid. Leaves two or three. Inflorescence unbranched with six to twelve non-resupinate flowers. Sepals and petals yellowish brown or brown, lip white with few red dots, becoming yellow with age. Lip adnate to basal fourth of the column, three-lobed, lateral lobes clasping column, roundish to oblong. Midlobe of lip with a crenate (edged with rounded tooth

FIGURE 8-5. *Prosthechea ochracea*. Drawing by Jane Herbst.

that points forward) midvein and sometimes two lateral veins. Callus oblong, distally three-toothed. Column teeth digitate fimbriate (lacerate) subequal. Capsule three-winged.

HABITAT AND DISTRIBUTION

México to Costa Rica. Found at 800–3000 meters in elevation, in a variety of habits.

FLOWERING TIME

Throughout year. Noted to be weedy.

CULTURE

Culture is typical for this genus, on the dry side in winter with more water in the summer months. The plants spread out nicely in a pot or on a flat surface, though they can be mounted.

COMMENT

Plate 92 shows a very pale color form; the normal color form is a brown-orange or a dull sulfur-yellow. No one submitted a slide of the normal form.

This is a nice plant if you get one of the varieties that open fully when in bloom, as the blooms are held well above the foliage and last a long time. Many varieties of this plant are cleistogamous, and do not open fully.

Eric Christenson (pers. comm., 2002) reports seeing a similar yet closely related, undescribed species in cultivation in Ecuador.

MEASUREMENTS

Pseudobulbs 3.5–9.5 cm long, 0.5–2 cm wide
Leaves 6–27 cm long, 0.4–1.7 cm wide
Inflorescence 4–15 cm long
Spathe 4 cm long
Sepals 3.8–12 mm long, 1.5–2.5 mm wide
Petals 3.6–5.5 mm long, 1.2–2 mm wide
Lip 3.2–13 6 mm long, 1–12 mm wide
Column 3–5 mm long

Prosthechea ortizii

Prosthechea ortizii (Dressler) W. E. Higgins. 1997. *Phytologia* 82 (5): 381. Basionym: *Encyclia ortizii* Dressler. 1995. *Novon* 5 (2): 140. Type: Costa Rica: Alajuela: Reserve San Ramón, approximately 30 kilometers NNE of San Ramón, 8–9 December 1984, *R. L. Dressler & Biología 350 288* (holotype: USJ).

DERIVATION OF NAME
Honors Rodolfo Ortiz, director of the San Ramon nature reserve.

DESCRIPTION
Plants grow in tufts. Pseudobulbs pear-shaped to oval. Leaf one. Inflorescence a raceme. Lip clawed (unguiculate), claw about 4 mm long, basally adnate to column, lip three-lobed, lateral lobes clasping column apex, midlobe at right angle to column. Callus prominent, triangular with laminar (platelike) high keel. Midtooth of column twice as long as lateral teeth.

HABITAT AND DISTRIBUTION
Costa Rica.

MEASUREMENTS
Pseudobulbs 3.7 cm long, 0.8–1.2 cm wide
Leaves 8.2–10 cm long, 2.3–2.6 cm wide
Inflorescence 12–24 cm long
Sepals 9 mm long, 2.8–3 mm wide
Petals 8 mm long, 2 mm wide
Lip 4 mm long
Column 7 mm long

Prosthechea panthera
PLATE 93

Prosthechea panthera (Reichenbach f.) W. E. Higgins. 1997. *Phytologia* 82 (5): 381. Basionym: *Epidendrum panthera* Reichenbach f. 1856. *Bonplandia* 4: 326.

SYNONYMS
Epidendrum papyriferum Schlechter. 1899. *Bulletin de l'Herbier Boissier* 7: 543.
Encyclia panthera (Reichenbach f.) Schlechter. 1918. *Beih. Bot. Centralbl.* 36 (2): 473.
Hormidium panthera (Reichenbach f.) Brieger. 1977. Schlechter's *Die Orchideen*, ed. 3, p. 574.

DERIVATION OF NAME
Latin *panthera*, "cougar" or "of the gods."

DESCRIPTION
Pseudobulbs 1–5 cm apart on rhizome, ovoid to spindle-shaped. Leaves two or three. Inflorescence unbranched with eight to fifteen non-resupinate flowers.

Sepals and petals olive green, blotched with dark red-brown becoming yellow or orange with age, lip white becoming yellow with age. Lip adnate to basal quarter of column, three-lobed, lateral lobes clasping column, subquadrate, midlobe quadrate-obdeltoid. Midlobe with three veins, midvein reaching apex. Callus narrowly oblong, sulcate, passing into a papillose or crenulate vein on midlobe. Column teeth digitate-fimbriate, subequal. Seed capsule is subcyclindric, not winged.

HABITAT AND DISTRIBUTION
Guatemala and México. Found at 1300–2000 meters in elevation, in dry oak and pine forests, also in somewhat wetter oak-liquidamber forest.

FLOWERING TIME
February to May.

MEASUREMENTS
Pseudobulbs 4–9 cm long, 1–1.5 cm wide
Leaves 8–22 cm long, 0.4–0.8 cm wide
Inflorescence 12–22 cm long
Spathe 2–4 cm long
Sepals 3.8–9 mm long, 1.5–3 mm wide
Petals 3.8–8 mm long, 1.2–2.5 mm wide
Lip 3.2–8 mm long, 2.5–4 mm wide
Column 3–6 mm long

Prosthechea serpentilingua

Prosthechea serpentilingua Withner & Hunt. 2001. *Orchid Digest* 65 (2): 79. Type: Brazil, unknown origin, probably the Organ Mountains. *Ex* cult. *David G. Hunt s.n.* (holotype: AMES). The type material consists of only two flowers from the same plant: one with the lip and column preserved in alcohol and glycerin mixture and the other with sepals, petals, and lip dissected, with the parts dried and mounted on a card.

DERIVATION OF NAME
After the tip of the lip, which protrudes from under the apex of the column as if it was a snake's tongue.

DESCRIPTION
The foliage is a glaucous green. The sepals and petals are a glaucous dull green to a greenish yellow. The lip has a yellowish cast toward the base. The somewhat

clear farinaceous coverings of the lip apex continue into a small patch on the lip claw. The column has a narrow base that swells slightly toward the tip to about 3 mm wide across the stigmatic area. The column wings or edges curl inward to form a groove, and each edge produces a wide auricle-like point that curls toward the stigma. The edges are terminated beyond in definite points as they reach the tip of the column. The tip of the column also has a toothed flaplike appendage that lies between the two pointed extensions of the edges. The column is greenish, and the column edges are a dull orange around the stigma.

HABITAT AND DISTRIBUTION
Brazil, possibly from the Organ Mountains.

COMMENT
The torsion of the lip and the in-turned edges of the column hold it tightly against the stigma, but it can be with some force fully pried away from the column for access to the nectary. It must require a creature of some strength and with a long tongue or proboscis to carry out pollination.

MEASUREMENTS
Dorsal sepal 15 mm long, 6 mm wide
Lateral sepals 13.5 mm long, 5.5 mm wide
Petals 14 mm long, 5.5 mm wide
Lip 12.5 mm long, 1.5 mm wide throughout its length and with no extended lateral lobes
Column 10 mm long
Pedicellate ovary 25 mm long

Prosthechea vitellina
FIGURE 8-6, PLATE 94

Prosthechea vitellina (Lindley) W. E. Higgins. 1997. *Phytologia* 82 (5): 381.
Basionym: *Epidendrum vitellinum* Lindley. 1831. *Gen. Sp. Orchid. Pl.* 97.

SYNONYMS
Epidendrum vitellinum var. *majus* Veitch. 1866. *Floral Mag.* 5: t. 261.
Epidendrum vitellinum var. *giganteum* Warner. 1887–1888. *Select Orch. Pl.* 3: t. 27.
Epidendrum vitellinum var. *autumnale* G. Wilson. 1913. *Orchid World* 4: 27.
Encyclia vitellina (Lindley) Dressler. 1961. *Brittonia* 13 (3): 265.
Hormidium vitellinum (Lindley) Brieger. 1977. Schlechter's *Die Orchideen*, ed. 3, p. 574.

DERIVATION OF NAME
Latin *vitellinus*, "dull yellow turning red," according to Lindley, referring to the flower color.

DESCRIPTION
Pseudobulbs glaucous (appearing to have powder on the surface), clustered, conic-ovoid, slightly flattened. Leaves one to three. Inflorescence simple or few branched, with four to twelve (more in cultivation) resupinate flowers. Sepals

FIGURE 8-6. *Prosthechea vitellina*. Drawing by Jane Herbst.

and petals vermilion, bright orange-red. Lip and column yellow or orange, anther tip and tip of lip orange-red. Lip basally adnate to column, unlobed, blade elliptic or oblong elliptic, acute, sides curled downward (revolute). Callus oblong, becoming three short veins on the blade. Column toothed, midtooth subquadrate, somewhat flattened, shorter than lateral teeth. Capsule triangular in cross section, ovary three-winged.

HABITAT AND DISTRIBUTION
El Salvador, Guatemala, and México; historically falsely reported from Panama. Found at 1500–2600 meters in elevation, in pine-oak forest, oak forest, cloud forest, and scrub forest on lava fields.

FLOWERING TIME
April to September.

CULTURE
Culture for this plant is intermediate to cool temperatures with a definite dry period in the winter and sparingly watered the rest of the year but always with high humidity and lots of light. Some people have no trouble growing this plant. Patricia struggles with it. Tom Perlite grows this species incredibly well, hanging it above his standard odontoglossums.

COMMENT
Here is another species that does not fit well anywhere. There are characters that fit with many genera, and yet never fit well enough. The column teeth seem to match the *Oestlundia*, but the size of the plants and the general stance of the flowers and the lack of the lip keels make this an unlikely relative of *Oestlundia*.

MEASUREMENTS
Pseudobulbs 2.5–6 cm long, 1.5–3 cm wide
Leaves 7–30 cm long, 0.8–5 cm wide
Inflorescence 12–45 cm long
Sepals 15–22 mm long, 3–8 mm wide
Petals 17–23 mm long, 6–11 mm wide
Lip 11.5–16 mm long, 3.5–5 mm wide
Column 6–7 mm long

CHAPTER 9

Miscellaneous Small Genera and Aberrant Species

Monotypic or even small genera with only two or three species remain enigmatic. Were they the result of mutant patterns that evolved and did not spread beyond the single species, or were the plants of these species the sole survivors of a once more commonly occurring genus? We will probably never know, but the mechanism of how such species evolved provides interesting contemplation. Some of these species have already been moved to their own genus; others that we mention we believe should be considered in the future as being in genera separate from *Epidendrum* or *Encyclia*.

Monotypic and Smaller Genera

Artorima, 278
Caularthron (Diacrium), 279
Dinema, 279
Encyclia kienastii, 280
Epidendropsis, 280
Epidendrum bracteolatum, 280
Epidendrum marmoratum, 281
Epidendrum stamfordianum, 281
Euchile, 282
Hagsatera, 282
Kalopternix, 282
Lanium, 283
Nidema, 283
Seraphyta, 283

Artorima

Artorima Dressler & Pollard. 1971. *Phytologia* 21: 439.

This fascinating genus is monotypic, comprising only one species *Artorima erubescens*. Dressler and Pollard published the genus in 1971 to accommodate the only insect foot-pollinated orchid found to date. The column is different from those otherwise found in *Epidendrum:* the thin edges have widened and extended toward each other across the lower surface of the column until they join with a slit in the midline. (Actually, this is the front or dorsal surface of the column from a developmental point of view.) This configuration encloses the stigmatic surface and produces a hollow chamber between the column and the base of the lip. When the pollinator's foot, with a pollinarium attached, slips through the slit in the region of the stigma, pollination is achieved. This is said to be done by bees that clamber around the flowers. The blossoms are indeed attractive to our eyes as well, with large panicles of rose-pink flowers, but the cool-growing plants are difficult to flower in cultivation. They may take several years, preferably in a basket or on a large section of vertical cork to accommodate the lengthy rhizomes, to become established enough to produce flowers (Withner 1988–2000, 5: 13). This singular species is found in the mountains of southern México and grows in the area known as the *tierra fria,* or cold earth.

Caularthron

Caularthron Rafinesque. 1836. *Flora Telluriana* 2: 40.

A subgenus from the Lindley key is *Diacrium* (Bentham. 1881. *Journ. Linn. Soc.* 18: 312.), a genus with a previous name, *Caularthron*. The species of this group have hollow homoblastic pseudobulbs, which provide nesting cavities for ants. They have a few apical leaves, grow on rocks near the seashore in full sunlight exposed to salt and spray or high in trees, and have lips that are fused to the column only at the base. The key character that separates the genus from *Epidendrum* is the unusual lip with two hornlike lateral lobes that are hollow and open onto the back of the lip. There are two, or possibly three, species in the genus (see Withner 1988–2000, vol. 5), and one species, *C. bilamellatum,* has a strong tendency toward cleistogamy (self-pollination without the opening of the flowers or the aid of a pollinator).

Dinema

Dinema Lindley. 1831. *Orch. Scel.* 16 (1826); *Gen. Sp. Orch. Pl.* 111.

Dinema polybulbon, the type species for this genus, has been moved around from one genus to another. The species was described by Swartz (1788) based on a specimen from Jamaica (*Nova Genera et Species Plantarum seu Prodromus* 124.) The plants are today commonly called both *Epidendrum* and *Encyclia polybulbon,* but we maintain the species as *Dinema*. The flowers of these low, creeping, pseudobulbous plants are not particularly decorative, but they differ significantly in column structure from other epidendrums. The column edges extend, at the apex of the column, into two long pointed structures that enclose the anther. An additional set of small points dips down on either side of the stigma. A good hand lens is needed to see such details on small flowers, but they are there. The column extensions as well as the anther cap are reddish in color. There are no lip lateral lobes, only thickenings, on the claw of the lip, and the lip midlobe is circular, white, and has a somewhat ruffled edge. The plants have been reported from México and Guatemala on the mainland, and the genus remains monotypic.

Encyclia kienastii

Encyclia kienastii (Reichenbach f.) Dressler & Pollard. 1971. *Phytologia* 21 (7): 437.

Encyclia kienastii, as fully described in Withner (1988–2000, 5:132) is a particularly rare entity from southern México. The flower is not similar to other species of *Encyclia* and is considered unique for its unusual qualities, which include its curved concave-backed column, its narrowly based lateral lobes protecting the callus, and flesh-pink-colored flowers with deeply colored veins. We feel that this is so unlike other species of *Encyclia* as to probably deserve separate generic status.

Epidendropsis

Epidendropsis Garay & Dunsterville. 1976. *Venezuelan Orchids Illustr.* 6: 39.

This is a small genus with *Epidendropsis violascens* the type species. Two additional species to date are *E. flexuosissima* and *E. vincentina*. Garay and Dunsterville published this genus and based it on a substantial key character difference from *Epidendrum*: the flowers of these species only produce two pollinia instead of the four pollinia typical of the genus *Epidendrum*. These are minute flowers on miniature plants of no horticultural use (Christenson, pers. comm.).

Epidendrum bracteolatum
PLATE 95

Epidendrum bracteolatum Presley. 1830. *Rel. Haenk.* i. 100.

Epidendrum bracteolatum grows on spiny acacia or cactus often in dry and barren chaparral areas of the coastal Ecuador. These habitats have low humidity, daytime sun, warmth, and nighttime condensation from dew or fog. The species, correctly known as *E. bracteolatum* Presley, is often encountered under its later synonym *E. collare* Lindley. The night-fragrant flowers are certainly *Epidendrum*-like with the lip fused to the column to its apex. The flowers are produced at the top of a raceme or few-branched panicle. The foliage is 2.5–7.5 cm long with leathery leaves at the apex of a thickened tough and fibrous stem.

Epidendrum marmoratum

Epidendrum marmoratum A. Richard & H. Galeotti. 1845. *Ann. Sc. Nat.*, ser. 3, 3: 21.

Epidendrum marmoratum has an unusual flattened flower stalk, unlike other *Encyclia* species. It produces its marbled flowers of white and maroon on smooth sheathed spindle-shaped pseudobulbs, unlike the ovoid or round pseudobulbs of *Encyclia*. It too is a unique species that probably deserves its own genus.

Epidendrum stamfordianum
PLATES 96, 97, 98

Epidendrum stamfordianum Bateman. 1838. *Orchid. México Guatemala* t. 11.

The key character that separates this species from *Epidendrum* proper is that the flower stalk arises from the rhizome (rootstock) of the plants on a separate leafless growth (called a radicle inflorescence by Lindley). It is a growth that is neither fleshy nor pseudobulbous, but it does have a series of somewhat overlapping elongated bracts along its length, and it terminates in the flower or flower cluster. The lip is fused to the column for about two-thirds of the column length and is distinctly three-lobed. The leaf-bearing stems of the plants are somewhat fleshy and spindle-shaped. There are no other species of *Epidendrum* like this one, though some of the same characteristics turn up occasionally in other genera. We can mention, for instance, the flowering of *Alamania punicea*, *Brassavola acaulis*, and *Cattleya walkeriana* on separate leafless growths.

Psilanthemum is a Lindley subgenus of *Epidendrum* originally proposed as a monotypic subgenus, having only one species, *E. stamfordianum*. The name was mentioned as being elevated to the generic level by Klotzsch in Stein's 1892 *Orchideenbuch* (p. 532), though no type species is given for reference, only that it is an *Epidendrum*. Our research has not found any reference that changes *Epidendrum stamfordianum* to *Psilanthemum*.

Euchile

Euchile (Dressler & G. E. Pollard) C. L. Withner. 1998. *Cattleyas and Their Relatives* 5: 137.

This genus with two species, *Euchile citrina* and *E. mariae*, has been a candidate for separation from *Epidendrum* for a while, and Withner (1998), using an earlier sectional name by Dressler and Pollard, raised it to generic status. The epithet means "beautiful lip" in reference to the large and expanded apex of the lip. There is little or no indication of lateral lobes. The plants of both species have a glaucous silvery gray appearance, and both species have large, rounded, and ribbed seedpods. The late Frederico Halbinger believed that three distinct species had been included in *E. citrina* (Christenson, pers. comm.) but never published his data. Having examined the data and specimens at the Kew Herbarium, Withner could not observe any distinctive species differences.

Hagsatera

Hagsatera Gonzalez Tamayo. 1974. *Orquídea (México)* 3 (11): 343.

This genus, endemic to México, has only two species, *Hagsatera brachycolumna* and *H. rosilloi*. The name honors Eric Hágsater who has done much with his field knowledge, cultivation, and conservation efforts. In addition, he has done much research on the orchids of México, established a national herbarium of Mexican orchids, and founded the journal *Orquídea*. Roberto Gonzalez Tamayo established the genus in 1974 with *H. brachycolumna* as type, noting that the flowers produced eight, not the four pollinia typical of *Epidendrum*.

Kalopternix

Kalopternix L. A. Garay & G. C. K. Dunsterville. 1976. *Venezuelan Orchids Illustr.* 6: 40.

This epithet of Greek origin indicates a plant with a shortened, prostrate, rhizomatous stem. The lip is attached to the column and the pollinia are completely different from those usually found in *Epidendrum*. Garay and Dunsterville in 1976 state, "the noncompressed pollinia . . . are attached to a pair of hyaline stipes." The genus includes three species: *K. deltoglossus, K. mantinianus,* and *K. sophronites.*

Lanium

Lanium Lindley ex Bentham. 1881. *Hook. Ic. Pl.* t. 1334.

Lanium is another of Lindley's subgenera elevated to the generic level. The key characters are scaly creeping stems with a small growth habit, small flowers, and the presence of woolly hairs on the backs of the sepals, the flower stalks, and the fruits. *Lanium* means woolly in Latin. Bentham established the genus in 1881 as distinct from *Epidendrum*. The lip is undivided and attached only at the base of the column. The pollinia are unique among these genera; the center two are larger, compressed, and the outer two are smaller, elongated and comma-shaped. The flowers are non-resupinate, and a bend in the flower stalk makes the flowers appear like little birds, if you have the right kind of imagination. The four species to date come from Brazil and adjacent territories. See Withner 1988–2000 (3: 90) for more details.

Nidema

Nidema Britton & Millspaugh. 1920. *Fl. Baham.* 94.

This genus was proposed by Britton and Millspaugh for a plant widely distributed from Cuba, Venezuela, and Suriname northwards through Panama to México and south to Ecuador: *Nidema ottonis*. This species had been described originally by Reichenbach as an *Epidendrum* in 1858 (*Hamburger Garten- und Blumenzeitung* 14: 213). Schlechter then in 1922 (*Repertorium Specierum Novarum Regni Vegetabilis, Beihefte* 17: 43) used the name of *Epidendrum boothii* var. *triandrum*, and in 1960 Schultes, followed by Dunsterville and Garay in 1966, restored it to *Epidendrum ottonis*, not wishing to use *Nidema* as the generic name.

The second species, *Nidema boothii*, has larger flowers, which are very fragrant, and is a nice addition to a mixed collection.

Nidema is an anagram of *Dinema*, although Britton and Millspaugh did not give any explanation why they removed the species from *Epidendrum*.

Seraphyta

Seraphyta Fischer & Meyer. 1840. *Bull. Sc. Acad. Petersb.* 8: 24.

This is another monotypic genus. Fischer and Meyer used the name *Seraphyta multiflora*. The type species is *Epidendrum diffusum*, found in Swartz's *Pro-*

dromus from 1788. Swartz's name *diffusum* has priority. The species now is *Seraphyta diffusa* Pfitzer ex Fawcett and Rendle. The small flowers of these plants have a short 3 mm column fused with the lip and many are produced in a cloudlike panicle reminiscent of a swarm of mosquitoes on a summer evening. The lip has no lateral lobes and is cordate to ovoid-shaped with the midnerve distinct and channeled.

Seraphyta diffusa (Swartz) Pfitzer ex Fawcett & Rendle. 1910. *Flora of Jamaica* 1: 81. Basionym: *Epidendrum diffusum* Swartz. 1788. *Nova Genera et Species Plantarum seu Prodromus* 121. Type: Jamaica, *Swartz s.n.* (holotype: S).

SYNONYM
Seraphyta multiflora Fischer & Meyer 1840. *Bull. Sc. Acad. Petersb.* 8: 24.

Selected References for Additional Reading

Acuña Galé, Julián Baldomero. 1938. *Catologo Descriptivo de las Orquideas Cubanas*. Bulletin 60 (June). Estacion Experimental Agronomica, Cuba.

Ames, Oakes, and Donovan Stewart Correll. 1952. *Orchids of Guatemala*. Fieldiana: Botany 26 (1). Chicago, Illinois: Field Museum of Natural History.

Ames, Oakes, and Donovan Stewart Correll. 1985. *Orchids of Guatemala and Belize*. New York: Dover Publications. 209–395.

Ames, Oakes, T. F. Hubbard, and C. Schweinfurth. 1936. *The Genus* Epidendrum *in the United States and Middle America*. Cambridge, Massachusetts: Botanical Museum, Harvard University.

Bennett, D. E., Jr., and E. A. Christenson. 1995. *Icones Orchidacearum Peruvianum*. Sarasota, Florida.

Benson, Lyman. 1959. *Plant Classification*. Massachusetts: D. C. Heath and Company.

Brieger, F., R. Maatsch, and K. Senghas. 1977. 3rd. ed. of R. Schlechter's *Die Orchideen*. Berlin and Hamburg: Verlag, Paul Parey. 509–576.

Carnevali, G., and I. Ramírez. 1988. Revisión del genero *Encyclia* para Venezuela. *XLVI Exposicíon de Orquideas, Comité de Orquideología* 3: 13–87.

Castro Neto, Vitorino Paiva , and Marcos Antonio Campacci. 2001. *Prosthechea carrii sp. nov. Orchid Digest* 65 (2): 74–75.

Castro Neto, Vitorino Paiva, and André Cardoso. 2003. Un nouvel *Encyclia* (Orchidaceae) de l'Amazonie brésilienne. *Richardiana* 3 (2): 69.

Chase, Mark W. 1986. A Reappraisal of the Oncidioid Orchids. *Systematic Botany* 11: 477–491.

Christenson, Eric A. 1996. MesoAmerican Orchid Studies VI: A Name Change for *Encyclia lancifolia. Lindleyana* 11 (4): 222.

Christenson, Eric A. 2002. Three Orchids from the Sacred City. *Orchids* 71(8): 714–716.

Cogniaux, A. 1893–1906. *Orchidaceae II, Tribus VII Laeliinae. Flora Brasiliensis (Martius)* 3 (5).

Dodson, Calaway H., and E. Hágsater. 1999. *Monographs in Systematic Botany. Missouri Botanical Garden* 75: 956.

Dodson, Calaway H., and David E. Bennett, Jr. 1989. *Icones Plantarum Tropicarum—Orchids of Peru,* Series II, Fascicle 1. St. Louis, Missouri: Missouri Botanical Garden.

Dodson, Calaway H., and Piedad Marmol de Dodson. 1989. *Icones Plantarum Tropicarum—Orchids of Ecuador,* Series 11, Fascicle 5. St. Louis, Missouri: Missouri Botanical Garden.

Dodson, Calaway H., and R. Vásquez Ch. 1989. *Icones Plantarum Tropicarum,* Series II, Fascicle 3 and 4. *Orchids of Bolivia.* St. Louis, Missouri: Missouri Botanical Garden.

Dressler, Robert L. 1961. A Reconsideration of the *Encyclia* (Orchidaceae). *Brittonia* 13: 253–266.

Dressler, Robert L. 1977. El Complejo de *Encyclia fragrans* en los Paises Andinos. *Orquideologia* 6 (4): 195–202.

Dressler, Robert L. 1981a. *Phylogeny and Classification of the Orchid Family.* Portland Oregon: Dioscorides Press.

Dressler, Robert L. 1981b. *The Orchids, Natural History and Classification.* Cambridge, Massachusetts: Harvard University Press.

Dressler, Robert L. 1984. The Delineation of the Genera in the *Epidendrum* Complex. *Orquídea (México)* 9 (2): 291–298.

Dressler, Robert L. 1993. *Field Guide to the Orchids of Costa Rica and Panama.* Ithaca and London: Comstock Publishing Associates.

Dressler, Robert L., and Glenn Pollard. 1971. Nomenclatural Notes on the Orchidaceae IV. *Phytologia* 21 (7): 438.

Dressler, Robert L., and Glenn Pollard. 1974. *The Genus* Encyclia *in México.* México, D.F.: Asociación Mexicana de Orquideología.

Dunsterville, G. C. K. and E. Dunsterville. 1980. *Epidendrum tigrinum* and *Epidendrum pamplonense*—Old Species Restored to Life. *American Orchid Society Bulletin* 49 (7): 716.

Dunsterville, G. C. K., and Leslie A. Garay. 1959–1976. *Venezuelan Orchids Illustrated,* 6 vols. London: Andre Deutsch.

Dunsterville, G. C. K., and Leslie A. Garay. 1979. *Orchids of Venezuela—an Illustrated Field Guide.* 3 vols. Cambridge, Massachusetts: Botanical Museum, Harvard University.

Escobar, Rodrigo. 1994–1998. *Native Colombian Orchids.* Medellín, Colombia: Compañía Litografica Nacional S.A. 2: 172–175; 5: 746–772.

Garay, Leslie A. 1971. Orquídeas Colombianas nuevas o críticas. *Orquideología* 6 (1): 16.

Gloudon, A., and C. Tobisch. 1995. *Orchids of Jamaica.* Jamaica, West Indies: The Press University of the West Indies. 49–58.

Hágsater Eric. 1993a. New Combinations in *Encyclia* and *Epidendrum. Orquídea (México)* 13 (1–2): 215–218.

Hágsater, Eric. 1993b. *Icones Orchidacearum* Fascicle 2 (plates 101–200), A Century of *Epidendrum.* México, D.F.: Asociación Mexicana de Orquideología.

Hágsater, Eric, and G. A. Salazar. 1990. *Icones Orchidacearum* Fascicle 1 (plates 1–100), *Orchids of México.* México, D.F.: Asociación Mexicana de Orquideología.

Hamer, Fritz. 1981. *Las Orquídeas de El Salvador.* San Salvador, El Salvador: Ministry of Education. 176–280.

Hamer, Fritz. 1982. *Icones Plantarum Tropicarum, Orchids of Nicaragua,* fascicle 13, part 1–6. Sarasota, Florida: Marie Selby Botanical Gardens.

Hamer, Fritz. 1988. Orchids of Central America. *Selbyana* 10: 184–235.

Higgins, W. E. 1997. A Reconsideration of the Genus *Prosthechea* (Orchidaceae). *Phytologia* 82 (5): 370–383.

Higgins, W. E. 1999. The Genus *Prosthechea*: An Old Name Resurrected. *Orchids* 68 (11): 1114.

Higgins, W. E. 2000. *Intergeneric and Intrageneric Phylogenetic Relationships of Encyclia (Orchidaceae) Based upon Holomorphology.* Ph.D. Thesis, Horticultural Sciences, University of Florida, Gainesville.

Higgins, W. E. 2001. *Oestlundia*: A New Genus of Orchidaceae in Laeliinae. *Selbyana* 22 (1): 1–4, f. 1–2.

Hoehne, F. C. 1947. *Arquivos de botanica do estado de São Paulo,* n.s., 2: 77–87.

Lindley, John. 1839. *Edwards's Botanical Register.*

Lindley, John. 1841. Notes upon the Genus *Epidendrum. Hooker's Journ. Bot.* 3: 81.

Lindley, John. 1847. *The Elements of Botany. . . .and a Glossary of Technical Terms.* London.

Lindley, John. 1853. *Folia Orchidacea* Epidendrum. London: Matthews.

Mayr, Hubert. 1998. *Orchid Names and Their Meanings.* Port Jervis, New Jersey: Lubrecht & Cramer.

McQueen, Jim, and Barbara McQueen. 1992. *Miniature Orchids.* Portland, Oregon: Timber Press. 76–81.

Miller, David, and Richard Warren. 1994. *Orchids of High Mountain Atlantic Rain Forest in SE Brazil.* Rio de Janeiro, Brazil: Salamandra Consultoria Editorial SA.

Northen, Rebecca Tyson. 1980. *Miniature Orchids.* New York: Van Nostrand Reinhold.

Northen, Rebecca Tyson. 1990. *Home Orchid Growing.* 4th ed. New York: Simon & Schuster.

Pabst, G. F. F., and E. Dungs. 1977. *Orchidaceae Brasilienses*, Band 1: 300–311. Hildeschein, Germany: Hagemann-Druck.

Pabst, G. F., and A. V. Pinto. 1981. An Attempt to Establish the Correct Statement for the Genus *Anacheilium* Hoffmgg. and a Revision of the Genus *Hormidium* Lindl. ex Heynh. *Bradea* 1 (23): 173. Published in Brazil in *Orquidéfilos e Orquidélogos*. Rio de Janeiro, by Expressão e Cultura.

Pridgeon, Alec M., P. J. Cribb, M. W. Chase, and F. N. Rasmussen. 1999. *Genera Orchidacearum*. vol. 1. Oxford, England: Oxford University Press.

Schweinfurth, Charles. 1959. *Orchids of Peru, Fieldiana: Botany* 30 (2). Chicago, Illinois: Chicago Natural History Museum. 91–164.

Senghas, Karlheinz. 2001. Namens wirrwarr mit und um *Encyclia*—mit einer neuen Art aus Peru. *Journal fur den Orchideenfreund* 8 (1): 58–64.

Small, John K. 1913. *Flora of Miami*. New York: Author.

Stearn, William T. 1995. *Botanical Latin*. Portland, Oregon: Timber Press.

Teuscher, Henry. 1969. Collectors Item: *Epidendrum prismatocarpum* var. *ionocentrum*. *American Orchid Society Bulletin* 38: 395–398.

Van den Berg, Cássio, W. E. Higgins, R. L. Dressler, Mark Whitten, Miguel A. Soto Arenas, Alastair Culham, and Mark W. Chase. 2000. A Phylogenetic Analysis of Laeliinae (Orchidaceae) Based on Sequence Data from Internal Transcribed Spacers (ITS) of Nuclear Ribosomal DNA. *Lindleyana* 15 (2): 96–114.

van der Pijl, L., and C. H. Dodson. 1966. *Orchid Flowers, Their Pollination and Evolution*. Miami: University Miami Press.

Vásquez, Roberto Ch., and Calaway H. Dodson. 1982. *Icones Plantarum Tropicarum*, ser. 1, fascicle 6. Sarasota, Florida: Marie Selby Botanical Gardens.

Wiard, Leon. 1987. *An Introduction to the Orchids of México*. Ithaca and London: Comstock Publishing Associates.

Williams, Louis O., and Paul H. Allen. 1980. *Orchids of Panama*. St. Louis, Missouri: Missouri Botanical Garden.

Withner, Carl. 1988–2000. *The Cattleyas and Their Relatives*. 6 vols. Portland, Oregon: Timber Press.

Withner, Carl. 2001. An Overview of the Genus *Prosthechea*. *Orchid Digest* 65 (2): 78–81.

Index of Persons

Acuña Galé, Ing. Julián Baldomero 29, 216
Allikas, Greg 258

Barker, George, Esq. 152
Bennett, David E., Jr. 51
Brasavole, Antonio Musa 209
Burmann, Johannes 28, 138
Bussey, Weyman 160

Calatayud, Gloria 135
Campacci, Marcos Antonio 61, 129
Campos Pôrto, Paulo de 60
Carr, George 61
Castro Neto, Victorino Paiva 61, 99
Christenson, Eric A. 58, 135
Christy, Thomas 261

Dressler, Robert L. 31, 207, 217, 263
Dunsterville, G. C. K. 59

Escobar, Gilberto 87

Farfan Rios, William 78
Faustus 80
Figura, Jim 87

García Castro, Joaquín B. 98
Garcia Esquivel, Carlos 85
Ghiesgbreght, Auguste Boniface 224
Greenwood, Ed 226

Hágsater, Eric 23, 282
Hajek, Carlos 84, 90, 91
Hajek, Frank 90
Halbinger, Frederico 282
Hartweg, Karl Theodor 92
Higgins, Wesley E. 23, 26, 137, 201, 207

Hoffmannsegg, Johann Centurius von 28
Hooker, William Joseph 148

Kautsky, Roberto 61, 99
König, Johann Gerland 141

Leonard, Bill 116
Linden, Jean 102
Lindley, John 14–15, 20–21, 23, 28, 29, 137, 138, 141, 148, 149, 151, 152, 201, 216, 217, 251, 279
Link, Johann Heinrich Friedrich 229
Linnaeus, Carl 28, 137

Mejia de Moreno, Esperanza 103
Moojen de Oliveira, J. 104

Ørsted, Anders Sandøe 145
Ortiz, Rodolfo 273
Östlund, Karl Erik Magnus 201

Parkinson, John 146
Peixoto, Mauro 59, 61
Perlite, Tom 277
Pineda Palacio, J. Bayron 87
Plumier, Charles 28
Pollard, Glen E. 201, 217
Portillo, José (Pepe) 54
Pupulin, Franco 207

Regnell, Anders 111

Small, John K. 149, 216
Swartz, Olaf 279

Vásquez Ch., Roberto 126

Withner, Carl 132

Index of Plant Names

Boldface indicates pages with main entries.
Italic indicates pages with color photos.

Alamania punicea 281
Amblostoma tridactylum, see *Prosthechea glauca*
Amphiglottis 29
Amphiglottium 21
Anacheilium 23, 24, 27, 28, 29, 30, 111, 152, 207, 217, 251
Anacheilium abbreviatum 26, 29, **40–42**, 45, 105, 124, *161*
Anacheilium aemulum 24, 26, 29, 31, **42–44**, 82, *161*
Anacheilium alagoense 29, 41, **45**, 124, *161*
Anacheilium allemanii 26, 29, **46–47**, *162*, *163*
Anacheilium allemanoides 29, **48**, *163*
Anacheilium aloisii 29, **49**, 135
Anacheilium baculus 29, **49–51**, 72, 100, *163*
Anacheilium bennettii 29, **51–52**, 83, *164*
Anacheilium brachychilum 29, **52–54**, 93, 111, *164*
Anacheilium bulbosum 28, 29, 48, **54–55**, 111, *165*
Anacheilium caetense 29, **55–56**, *166*
Anacheilium calamarium 26, 29, **57–59**, 115, *166*
Anacheilium campos-portoi 29, **59–61**, 99, *167*
Anacheilium carrii 29, **61–62**, *167*
Anacheilium Calatayud 513 **135**
Anacheilium chacaoense 29, **63–65**, 67, 93, 100, *167*
Anacheilium chimborazoense 24, 29, **65–67**, *168*
Anacheilium chondylobulbon 29, **67–69**, *169*
Anacheilium cochleatum 17, 24, 26, 28, 29, **69–71**, 82, 122, *170*
Anacheilium cochleatum var. *triandrum* **71–72**
Anacheilium confusum 29, **72–73**

Anacheilium crassilabium 29, **74–77**, 106, 121, *170, 171*
Anacheilium faresianum 29, **77–78**, *171*
Anacheilium farfanii 29, **78–79**
Anacheilium faustum 26, 29, 31, **80–81**, 151, *171*
Anacheilium fragrans 24, 28, 29, 31, 41, 65, **81–82**, 96, 136, 151, *172*
Anacheilium fuscum 29, **83–84**, 93, *172*
Anacheilium garcianum 29, **84–86**, *172*
Anacheilium gilbertoi 29, **86–87**, *173*
Anacheilium glumaceum 29, 80, 81, **87–89**, 151, *174*
Anacheilium grammatoglossum, see *Hormidium grammatoglossum*
Anacheilium hajekii 29, 79, **89–91**, 127, *174*
Anacheilium hartwegii 29, 83, **91–93**, 127, *174*
Anacheilium ionophlebium 29, 64, **93–94**, 124, *175*
Anacheilium janeirense 29, **94–96**, *175*
Anacheilium jauanum **96–97**
Anacheilium joaquingarcianum **98**
Anacheilium kautskyi 61, **98–99**, *175*
Anacheilium lambda 26, 87, **99–100**
Anacheilium lindenii **101–102**, *176*
Anacheilium lividum, see *Pollardia livida*
Anacheilium mejia **102–103**, *176*
Anacheilium moojenii 26, **104–105**
Anacheilium neurosum **105**
Anacheilium pamplonense 77, **106**
Anacheilium papilio **107–108**, 120, 133, 151, *177*
Anacheilium punctiferum, see *A. calamarium*
Anacheilium radiatum 76, **108–110**, 124, *177*
Anacheilium regnellianum **110–111**
Anacheilium santanderense **111–113**, *178*
Anacheilium sceptrum **113–115**, *178*

Index of Plant Names   291

Anacheilium sessiliflorum 115–116
Anacheilium simum 116–117, *178*
Anacheilium spondiadum 118–119, *179*
Anacheilium suzanense 26, 119–120
Anacheilium tigrinum 77, 106, **120–122**, *179*
Anacheilium trulla 122–123, *179*
Anacheilium undescribed species 135
Anacheilium undescribed species from Peru 133–134
Anacheilium vagans 41, 45, **124–125**, *179*
Anacheilium vasquezii **126–127**, *180*
Anacheilium venezuelanum 26, **127–128**, *180*
Anacheilium vespa 55, 77, 106, 120, **128–130**, 151, *181*
Anacheilium vinaceum **130–131**, 134
Anacheilium vita 114, **131–132**, *181*
Anacheilium widgrenii 26, 31, 120, **132–133**, *181*
Artorima 15, 149, **278–279**
Artorima erubescens 278
Auliza 29, 137
Auliza ciliaris, see *Coilostylis ciliaris*
Auliza clavatum, see *Coilostylis clavata*
Auliza lacertinum, see *Coilostylis lacertina*
Auliza wilsoni, see *Anacheilium crassilabium*
Aulizeum 23, 29, 137, 151
Aulizeum cochleatum, see *Anacheilium cochleatum*
Aulizeum falcatum, see *Coilostylis falcata*
Aulizeum glumaceum, see *Anacheilium glumaceum*
Aulizeum tigrinum, see *Anacheilium tigrinum*
Aulizeum variegatum, see *Anacheilium crassilabium*
Aulizodium 137

Barkeria naevosa 151
Brassavola acaulis 281
Brassavola suaveolens, see *Coilostylis falcata*
Broughtonia 150
Broughtonia aurea, see *Cattleya aurantiaca*

Cattleya 16, 27
Cattleya aurantiaca 151, 216
Cattleya bowringiana 136
Cattleya guatemalensis 16
Cattleya hardyana 16
Cattleya walkeriana 281
Caularthron **279**
Caularthron bilamellatum 279
Cochleata 28

Coelogyne triptera, see *Hormidium pygmaeum*
Coilostylis 23, 137–139
Coilostylis ciliaris **139–140**, 142, 143, 145, *183*
Coilostylis clavatum **141–142**, *183*
Coilostylis cuspidata **142–143**
Coilostylis falcata **143–144**, *184*
Coilostylis lacertina **144**
Coilostylis oerstedii **145**, *184*
Coilostylis parkinsoniana 143, **145–146**, *185*
Coilostylis vivipara **146–147**, *185*
Cymbidium aloifolium 146

Diacrium 150, 279
Diacrium bidentatum, see *Prosthechea boothiana*
Diacrium bilamellatum, see *Caularthron bilamellatum*
Dichaea glauca 252
Didothion clavatum, see *Coilostylis clavata*
Dimerandra 15, 149
Dinema 15, 149, **279–280**
Dinema polybulbon 279
Dipodium 141
Dipodium squamatum 260

Encyclia 18, 23, 24, 27, 28, 29, 148–151, 201, 217, 251, 278
Encyclia abbreviata, see *Anacheilium abbreviatum*
Encyclia aemula, see *Anacheilium aemulum*
Encyclia alagoensis, see *Anacheilium alagoense*
Encyclia allemanii, see *Anacheilium allemanii*
Encyclia allemanoides, see *Anacheilium allemanoides*
Encyclia almasii, see *Anacheilium glumaceum*
Encyclia amabilis (Linden & Reichenbach f.) Schlechter, see *Pollardia concolor*
Encyclia amabilis Schlechter, see *Pollardia michuacana*
Encyclia arminii, see *Hormidium arminii*
Encyclia aurea, see *Cattleya aurantiaca*
Encyclia baculus, see *Anacheilium baculus*
Encyclia bennettii, see *Anacheilium bennettii*
Encyclia bicamerata, see *Prosthechea bicamerata*
Encyclia bidentata, see *Prosthechea boothiana*
Encyclia boothiana, see *Prosthechea boothiana*
Encyclia boothiana subsp. *erythronioides*, see *Prosthechea boothiana*
Encyclia boothiana subsp. *favoris*, see *Prosthechea boothiana*
Encyclia brachiata, see *Prosthechea brachiata*

Encyclia brachychila, see *Anacheilium brachychilum*
Encyclia bracteata 150
Encyclia brassavolae, see *Panarica brassavolae*
Encyclia bulbosa, see *Anacheilium bulbosum*
Encyclia caetense, see *Anacheilium caetense*
Encyclia caetensis, see *Anacheilium caetense*
Encyclia calamaria, see *Anacheilium calamarium*
Encyclia campos-portoi, see *Anacheilium camposportoi*
Encyclia campylostalix, see *Pollardia campylostalix*
Encyclia candollei 151
Encyclia chacaoensis, see *Anacheilium chacaoense*
Encyclia chimborazoensis, see *Anacheilium chimborazoense*
Encyclia chiriquensis, see *Pollardia varicosa*
Encyclia chondylobulbon, see *Anacheilium chondylobulbon*
Encyclia cochleata, see *Anacheilium cochleatum*
Encyclia cochleata subsp. *triandra*, see *Anacheilium cochleatum*
Encyclia cochleata var. *triandra* f. *albidoflava*, see *Anacheilium cochleatum*
Encyclia concolor, see *Pollardia concolor*
Encyclia crassilabia, see *Anacheilium crassilabium*
Encyclia cretacea, see *Prosthechea cretacea*
Encyclia cyanocolumna, see *Oestlundia cyanocolumna*
Encyclia dasilvae 148
Encyclia davidhuntii 151
Encyclia deamii, see *Pollardia livida*
Encyclia dickdentii 151
Encyclia distantiflora, see *Oestlundia distantiflora*
Encyclia faresiana, see *Anacheilium faresianum*
Encyclia fausta, see *Anacheilium faustum*
Encyclia fortunae, see *Prosthechea fortunae*
Encyclia fragrans, see *Anacheilium fragrans*
Encyclia fragrans subsp. *aemula*, see *Anacheilium aemulum*
Encyclia fragrans var. *aemulum*, see *Anacheilium aemulum*
Encyclia fusca, see *Anacheilium fuscum*
Encyclia garciana, see *Anacheilium garcianum*
Encyclia ghiesbreghtiana, see *Pollardia ghiesbreghtiana*
Encyclia gilbertoi, see *Anacheilium gilbertoi*
Encyclia glauca, see *Prosthechea glauca*
Encyclia glumacea, see *Anacheilium glumaceum*
Encyclia grammatoglossa, see *Hormidium grammatoglossum*
Encyclia greenwoodiana, see *Pollardia greenwoodiana*
Encyclia guttata, see *Prosthechea guttata*
Encyclia hartwegii, see *Anacheilium hartwegii*
Encyclia hastata, see *Pollardia hastata*
Encyclia hoehnei, see *Prosthechea christyana*
Encyclia icthyphylla, see *Pollardia michuacana*
Encyclia inversa, see *Anacheilium bulbosum*
Encyclia ionocentra, see *Panarica ionocentra*
Encyclia ionophlebia, see *Anacheilium ionophlebium*
Encyclia jauana, see *Anacheilium jauanum*
Encyclia kautskyi, see *Anacheilium kautskyi*
Encyclia kienastii **280**
Encyclia kundergraberi 151
Encyclia lambda, see *Anacheilium lambda*
Encyclia lancifolia, see *Anacheilium cochleatum*
Encyclia leopardina, see *Anacheilium crassilabium*
Encyclia limbata, see *Prosthechea glauca*
Encyclia linearis, see *Oestlundia luteorosea*
Encyclia linearloba 151
Encyclia linkiana, see *Pollardia linkiana*
Encyclia livida, see *Pollardia livida*
Encyclia luteorosea, see *Oestlundia luteorosea*
Encyclia maculosa, see *Prosthechea guttata*
Encyclia magnispatha, see *Prosthechea magnispatha*
Encyclia mapiriensis 151
Encyclia megahybos, see *Prosthechea christyana*
Encyclia michuacana, see *Pollardia michuacana*
Encyclia moojenii, see *Anacheilium moojenii*
Encyclia naevosa, see *Barkeria naevosa*
Encyclia neurosa, see *Anacheilium neurosum*
Encyclia obpiribulbon, see *Pollardia obpiribulbon*
Encyclia ochracea, see *Prosthechea ochracea*
Encyclia organensis see *Anacheilium calamarium*
Encyclia ortizii, see *Prosthechea ortizii*
Encyclia pachyantha 148
Encyclia pamplonensis, see *Anacheilium pamplonense*
Encyclia panthera, see *Prosthechea panthera*
Encyclia papilio, see *Anacheilium papilio*
Encyclia paraensis 148
Encyclia pastoris **250**
Encyclia pentotis, see *Anacheilium baculus*
Encyclia pipio, see *Anacheilium calamarium*

Index of Plant Names   293

Encyclia polybulbon, see *Dinema polybulbon*
Encyclia pringlei, see *Pollardia pringlei*
Encyclia prismatocarpa, see *Panarica prismatocarpa*
Encyclia pruinosa, see *Pollardia concolor*
Encyclia pseudopygmaea, see *Hormidium pseudopygmaeum*
Encyclia pterocarpa, see *Pollardia pterocarpa*
Encyclia pulcherrima, see *Anacheilium vinaceum*
Encyclia punctifera see *Anacheilium calamarium*
Encyclia pygmaea, see *Hormidium pygmaeum*
Encyclia racemifera, see *Hormidium racemiferum*
Encyclia radiata, see *Anacheilium radiatum*
Encyclia regnelliana, see *Anacheilium regnellianum*
Encyclia rhombilabia, see *Pollardia rhombilabia*
Encyclia rhynchophora, see *Hormidium rhynchophorum*
Encyclia sceptra, see *Anacheilium sceptrum*
Encyclia semiaptera, see *Pollardia semiaptera*
Encyclia sessiliflora, see *Anacheilium sessiliflorum*
Encyclia sima, see *Anacheilium simum*
Encyclia spondiada, see *Anacheilium spondiadum*
Encyclia squamata, see *Prosthechea christyana*
Encyclia subaquilum 151
Encyclia suzanensis, see *Anacheilium suzanense*
Encyclia tampensis 149
Encyclia tenuissima, see *Oestlundia tenuissima*
Encyclia tessellata, see *Pollardia livida*
Encyclia tigrina, see *Anacheilium tigrinum*
Encyclia trautmannii **135–136**
Encyclia triptera, see *Hormidium pygmaeum*
Encyclia tripunctata, see *Pollardia tripunctata*
Encyclia trulla, see *Anacheilium trulla*
Encyclia vagans, see *Anacheilium vagans*
Encyclia varicosa, see *Pollardia varicosa*
Encyclia varicosa subsp. *leiobulbon*, see *Pollardia varicosa*
Encyclia venezuelana, see *Anacheilium venezuelanum*
Encyclia venosa, see *Pollardia venosa*
Encyclia vespa, see *Anacheilium crassilabium*
Encyclia virgata, see *Pollardia michuacana*
Encyclia viridiflora 148, 149
Encyclia vitellina, see *Prosthechea vitellina*
Encyclia wendlandiana, see *Pollardia venosa*

Encyclia widgrenii, see *Anacheilium widgrenii*
Encyclium 20
Encyclium brassavolae, see *Panarica brassavolae*
Epicattleya (Epc.) Francis Dyer 136, *182*
Epicladium 29, 216, 217
Epicladium boothianum, see *Prosthechea boothiana*
Epidendropsis 15, **280**
Epidendropsis flexuosissima 280
Epidendropsis vincentina 280
Epidendropsis violascens 280
Epidendrum 14, 15, 20–21, 27, 28, 137, 148–151, 201, 207, 216–217, 251, 278, 280, 282
Epidendrum abbreviatum, see *Anacheilium abbreviatum*
Epidendrum acuminatum, see *Anacheilium baculus*
Epidendrum aemulum, see *Anacheilium aemulum*
Epidendrum aemulum var. *brevistriatum*, see *Anacheilium aemulum*
Epidendrum alagoense, see *Anacheilium alagoense*
Epidendrum allemanii, see *Anacheilium allemanii*
Epidendrum allemanoides, see *Anacheilium allemanoides*
Epidendrum ?allemanoides , see *Anacheilium allemanoides*
Epidendrum almasii, see *Anacheilium glumaceum*
Epidendrum aloifolium Bateman, see *Coilostylis parkinsoniana*
Epidendrum aloifolium Linnaeus, see *Cymbidium aloifolium*
Epidendrum aloisii, see *Anacheilium aloisii*
Epidendrum amabile, see *Pollardia concolor*
Epidendrum apiculatum, see *Anacheilium aemulum*
Epidendrum arminii, see *Hormidium arminii*
Epidendrum articulatum, see *Pollardia livida*
Epidendrum aurantiacum, see *Cattleya aurantiaca*
Epidendrum aureum, see *Cattleya aurantiaca*
Epidendrum auriculigerum, see *Panarica ionocentra*
Epidendrum auritum, see *Nidema boothii*
Epidendrum baculibulbum, see *Anacheilium crassilabium*

Epidendrum baculus, see *Anacheilium baculus*
Epidendrum beyrodtianum, see *Anacheilium baculus*
Epidendrum bicameratum, see *Prosthechea bicamerata*
Epidendrum bidentatum, see *Prosthechea boothiana*
Epidendrum bifidum, see *Anacheilium papilio*
Epidendrum boothianum, see *Prosthechea boothiana*
Epidendrum boothianum var. *erythronioides*, see *Prosthechea boothiana*
Epidendrum boothii var. *triandrum*, see *Nidema ottonis*
Epidendrum brachiatum, see *Prosthechea brachiata*
Epidendrum brachychilum, see *Anacheilium brachychilum*
Epidendrum bracteolatum **280–281**
Epidendrum brassavolae, see *Panarica brassavolae*
Epidendrum brevicolumna, see *Hagsatera*
Epidendrum bulbosum, see *Anacheilium bulbosum*
Epidendrum caespitosum, see *Hormidium pygmaeum*
Epidendrum calamarium, see *Anacheilium calamarium*
Epidendrum calamarium var. *brevifolium*, see *Anacheilium calamarium*
Epidendrum calamarium var. *latifolium*, see *Anacheilium calamarium*
Epidendrum calamarium var. *longifolium*, see *Anacheilium calamarium*
Epidendrum campylostalix, see *Pollardia campylostalix*
Epidendrum chacaoensis, see *Anacheilium chacaoense*
Epidendrum chimborazoensis, see *Anacheilium chimborazoense*
Epidendrum chiriquense, see *Pollardia varicosa*
Epidendrum chondylobulbon, see *Anacheilium chondylobulbon*
Epidendrum christii, see *Anacheilium crassilabium*
Epidendrum christyanum, see *Prosthechea christyana*
Epidendrum ciliare, see *Coilostylis ciliaris*
Epidendrum ciliare var. *cuspidatum*, see *Coilostylis cuspidata*
Epidendrum ciliare var. *minor*, see *Coilostylis ciliaris*
Epidendrum ciliare var. *oerstedii*, see *Coilostylis oerstedii*
Epidendrum ciliare var. *squamatum*, see *Coilostylis ciliaris*
Epidendrum ciliare var. *viscidium*, see *Coilostylis ciliaris*
Epidendrum cinnamomeum, see *Pollardia pterocarpa*
Epidendrum clavatum, see *Coilostylis clavata*
Epidendrum clavatum var. *purpurascens*, see *Coilostylis clavata*
Epidendrum cochleatum, see *Anacheilium cochleatum*
Epidendrum cochleatum var. *costaricense*, see *Anacheilium cochleatum*
Epidendrum cochleatum var. *pallidum*, see *Anacheilium cochleatum*
Epidendrum cochleatum var. *triandrum*, see *Anacheilium cochleatum*
Epidendrum collare, see *Epidendrum bracteolatum*
Epidendrum concolor, see *Pollardia concolor*
Epidendrum condylochilum, see *Pollardia livida*
Epidendrum confusum, see *Anacheilium baculus*
Epidendrum cordatum, see *Anacheilium aemulum*
Epidendrum coriaceum, see *Anacheilium crassilabium*
Epidendrum costaricense, see *Coilostylis oerstedii*
Epidendrum crassilabium, see *Anacheilium crassilabium*
Epidendrum cuspidatum, see *Coilostylis cuspidata*
Epidendrum cuspidatum var. *brachysepalum*, see *Coilostylis ciliaris*
Epidendrum cyanocolumnum, see *Oestlundia cyanocolumna*
Epidendrum dasytaenia, see *Pollardia livida*
Epidendrum deamii, see *Pollardia livida*
Epidendrum diffusum, see *Seraphyta diffusa*
Epidendrum diguetii, see *Pollardia tripunctata*
Epidendrum distantiflorum, see *Oestlundia distantiflora*
Epidendrum ensicaulon, see *Pollardia venosa*
Epidendrum erythronioides, see *Prosthechea boothiana*
Epidendrum falcatum, see *Coilostylis falcata*

Epidendrum falcatum var. *zeledoniae*, see *Coilostylis parkinsoniana*
Epidendrum fallax, see *Anacheilium lindenii*
Epidendrum fallax var. *flavescens*, see *Anacheilium lindenii*
Epidendrum faustum, see *Anacheilium faustum*
Epidendrum favoris, see *Prosthechea boothiana*
Epidendrum feddeanum, see *Anacheilium crassilabium*
Epidendrum flabellatum, see *Encyclia candollei*
Epidendrum flavescens, see *Anacheilium lindenii*
Epidendrum foliis subradicalibus, lanceolatis..., see *Anacheilium papilio*
Epidendrum foliis subradicalibus, oblongis..., see *Coilostylis ciliaris*
Epidendrum fragrans, see *Anacheilium fragrans*
Epidendrum fragrans var. *aemulum*, see *Anacheilium aemulum*
Epidendrum fragrans var. *alticallum*, see *Anacheilium aemulum*
Epidendrum fragrans var. *brevistriatum*, see *Anacheilium aemulum*
Epidendrum fragrans var. *ionoleucum*, see *Anacheilium fragrans*
Epidendrum fragrans var. *janeirense*, see *Anacheilium aemulum*
Epidendrum fragrans var. *magnum*, see *Anacheilium fragrans*
Epidendrum fragrans var. *megalanthum*, see *Anacheilium baculus*
Epidendrum fragrans var. *micranthum*, see *Anacheilium aemulum*
Epidendrum fragrans var. *pachypus*, see *Anacheilium fragrans*
Epidendrum fragrans var. *rivularium*, see *Anacheilium aemulum*
Epidendrum fuscum, see *Anacheilium fuscum*
Epidendrum garcianum, see *Anacheilium garcianum*
Epidendrum ghiesbreghtianum, see *Pollardia ghiesbreghtiana*
Epidendrum gilbertoi, see *Anacheilium gilbertoi*
Epidendrum glaucovirens, see *Prosthechea glauca*
Epidendrum glaucum 252
Epidendrum glumaceum, see *Anacheilium glumaceum*
Epidendrum glumibracteatum, see *Coilostylis clavata*
Epidendrum grammatoglossum, see *Hormidium grammatoglossum*
Epidendrum guttatum, see *Prosthechea guttata*
Epidendrum hartwegii, see *Anacheilium hartwegii*
Epidendrum hastatum, see *Pollardia hastata*
Epidendrum henrici, see *Pollardia livida*
Epidendrum hoehnei, see *Prosthechea christyana*
Epidendrum hoffmanii, see *Anacheilium ionophlebium*
Epidendrum icthyphyllum, see *Pollardia michuacana*
Epidendrum indusiatum, see *Coilostylis lacertina*
Epidendrum inversum, see *Anacheilium bulbosum*
Epidendrum ionocentrum, see *Panarica ionocentra*
Epidendrum ionoleucum, see *Anacheilium fragrans*
Epidendrum ionophlebium, see *Anacheilium ionophlebium*
Epidendrum kraenzlinii, see *Anacheilium papilio*
Epidendrum krugii, see *Anacheilium papilio*
Epidendrum lacertinum, see *Coilostylis lacertina*
Epidendrum lactiflorum, see *Coilostylis falcata*
Epidendrum lambda, see *Anacheilium lambda*
Epidendrum lancifolium, see *Anacheilium cochleatum*
Epidendrum langlassei, see *Anacheilium trulla*
Epidendrum latro, see *Anacheilium bulbosum*
Epidendrum leiobulbon, see *Pollardia varicosa*
Epidendrum leopardinum, see *Anacheilium crassilabium*
Epidendrum ligulatum 151
Epidendrum limbatum, see *Prosthechea glauca*
Epidendrum lindenii, see *Anacheilium lindenii*
Epidendrum lineare, see *Oestlundia luteorosea*
Epidendrum lineatum 205
Epidendrum linkianum, see *Pollardia linkiana*
Epidendrum lividum, see *Pollardia livida*
Epidendrum longiflorum 138
Epidendrum longipes, see *Anacheilium crassilabium*
Epidendrum lunaeanum, see *Pollardia varicosa*
Epidendrum luteoroseum, see *Oestlundia luteorosea*
Epidendrum luteum, see *Coilostylis ciliaris*
Epidendrum macrothyrsoides, see *Anacheilium sceptrum*

Epidendrum maculatum, see *Panarica prismatocarpa*
Epidendrum maculosum, see *Prosthechea guttata*
Epidendrum madrense, see *Anacheilium ionophlebium*
Epidendrum magnispathum, see *Prosthechea magnispatha*
Epidendrum marginatum, see *Anacheilium radiatum*
Epidendrum marmoratum 281
Epidendrum megahybos, see *Prosthechea christyana*
Epidendrum michuacanum, see *Pollardia michuacana*
Epidendrum micropus, see *Pollardia tripunctata*
Epidendrum miserum 152
Epidendrum monanthum, see *Hormidium pygmaeum*
Epidendrum moojenii, see *Anacheilium moojenii*
Epidendrum naevosum, see *Barkeria naevosa*
Epidendrum neurosum, see *Anacheilium neurosum*
Epidendrum nocturnum 22
Epidendrum ochraceum, see *Prosthechea ochracea*
Epidendrum oerstedii, see *Coilostylis oerstedii*
Epidendrum organense, see *Anacheilium calamarium*
Epidendrum ottonis, see *Nidema ottonis*
Epidendrum pachyanthum, see *Anacheilium brachychilum*
Epidendrum pachycarpum, see *Anacheilium chacaoense*
Epidendrum pachysepalum, see *Anacheilium crassilabium*
Epidendrum pamplonense, see *Anacheilium pamplonense*
Epidendrum panthera, see *Prosthechea panthera*
Epidendrum papilio, see *Anacheilium papilio*
Epidendrum papilionaceum, see *Anacheilium papilio*
Epidendrum papilionaceum var. *grandiflorum*, see *Anacheilium papilio*
Epidendrum papyriferum, see *Prosthechea panthera*
Epidendrum parkinsonianum, see *Coilostylis parkinsoniana*
Epidendrum parkinsonianum var. *falcatum*, see *Coilostylis falcata*

Epidendrum parviflorum, see *Prosthechea ochracea*
Epidendrum pastoris Link & Otto, see *Pollardia linkiana*
Epidendrum pastoris Llave & Lexarza, see *Encyclia pastoris*
Epidendrum pentotis, see *Anacheilium baculus*
Epidendrum phymatoglossum, see *Pollardia varicosa*
Epidendrum pipio, see *Anacheilium calamarium*
Epidendrum playcardium, see *Anacheilium spondiadum*
Epidendrum polybulbon, see *Dinema polybulbon*
Epidendrum pringlei, see *Pollardia pringlei*
Epidendrum prismatocarpum, see *Panarica prismatocarpa*
Epidendrum prismatocarpum var. *ionocentrum*, see *Panarica ionocentra*
Epidendrum prorepens, see *Anacheilium abbreviatum*
Epidendrum pruinosum, see *Pollardia concolor*
Epidendrum psilanthemum, see *Coilostylis clavata*
Epidendrum pterocarpum, see *Pollardia pterocarpa*
Epidendrum pulcherrimum 130, 134
Epidendrum punctiferum, see *Anacheilium calamarium*
Epidendrum punctulatum, see *Pollardia concolor*
Epidendrum purpurascens, see *Coilostylis clavata*
Epidendrum pygmaeum, see *Hormidium pygmaeum*
Epidendrum quadratum, see *Pollardia varicosa*
Epidendrum quadridentatum, see *Hormidium grammatoglossum*
Epidendrum radiatum, see *Anacheilium radiatum*
Epidendrum ramirezzi, see *Pollardia varicosa*
Epidendrum regnellianum, see *Anacheilium regnellianum*
Epidendrum rhabdobulbon, see *Anacheilium crassilabium*
Epidendrum rhopalobulbon, see *Anacheilium crassilabium*
Epidendrum rhynchophorum, see *Hormidium rhynchophorum*
Epidendrum rueckerae, see *Anacheilium lambda*
Epidendrum saccharatum, see *Anacheilium crassilabium*

Epidendrum sceptrum, see *Anacheilium sceptrum*
Epidendrum seriatum, see *Oestlundia luteorosea*
Epidendrum sessiliflorum, see *Anacheilium sessiliflorum*
Epidendrum sphenoglossum, see *Anacheilium sceptrum*
Epidendrum spondiadum, see *Anacheilium spondiadum*
Epidendrum squalidum 151
Epidendrum squamatum, see *Prosthechea christyana*
Epidendrum stamfordianum 199, 200, **281**
Epidendrum subaquilum, see *Encyclia subaquilum*
Epidendrum subulatifolium 201
Epidendrum susanense, see *Anacheilium suzanense*
Epidendrum suzanense, see *Anacheilium suzanense*
Epidendrum tampense, see *Encyclia tampensis*
Epidendrum tenuissimum, see *Oestlundia tenuissima*
Epidendrum tessellatum, see *Pollardia livida*
Epidendrum tigrinum, see *Anacheilium tigrinum*
Epidendrum triandrum, see *Anacheilium cochleatum*
Epidendrum tripterum, see *Pollardia semiaptera*
Epidendrum tripunctatum, see *Pollardia tripunctata*
Epidendrum triste, see *Prosthechea ochracea*
Epidendrum trulla, see *Anacheilium trulla*
Epidendrum umlaufii, see *Coilostylis oerstedii*
Epidendrum uniflorum, see *Hormidium pygmaeum*
Epidendrum uro-skinneri, see *Panarica prismatocarpa*
Epidendrum vagans, see *Anacheilium vagans*
Epidendrum vaginatum, see *Anacheilium fragrans*
Epidendrum varicosum, see *Pollardia varicosa*
Epidendrum variegatum, see *Anacheilium crassilabium*
Epidendrum variegatum var. *angustipetalum,* see *Anacheilium crassilabium*
Epidendrum variegatum var. *coriaceum,* see *Anacheilium crassilabium*
Epidendrum variegatum var. *crassilabium,* see *Anacheilium crassilabium*

Epidendrum variegatum var. *leopardinum,* see *Anacheilium crassilabium*
Epidendrum variegatum var. *lineatum,* see *Anacheilium crassilabium*
Epidendrum variegatum var. *virens,* see *Anacheilium crassilabium*
Epidendrum venezuelanum, see *Anacheilium venezuelanum*
Epidendrum venosum, see *Pollardia venosa*
Epidendrum vespa, see *Anacheilium vespa*
Epidendrum virgatum, see *Pollardia michuacana*
Epidendrum virgatum var. *pallens,* see *Pollardia michuacana*
Epidendrum viscidum, see *Coilostylis ciliaris*
Epidendrum vitellinum, see *Prosthechea vitellina*
Epidendrum vitellinum var. *autumnale,* see *Prosthechea vitellina*
Epidendrum vitellinum var. *giganteum,* see *Prosthechea vitellina*
Epidendrum vitellinum var. *majus,* see *Prosthechea vitellina*
Epidendrum viviparum, see *Coilostylis vivipara*
Epidendrum wendlandianum, see *Pollardia venosa*
Epidendrum widgrenii, see *Anacheilium widgrenii*
Epithecia glauca, see *Prosthechea glauca* 252
Euchile 15, 149, **282**
Euchile citrina 282
Euchile mariae 282
Euepidendrum 21

Glumacea 28, 30, 31

Hagsatera 15, **282**
Hagsatera brachycolumna 282
Hagsatera rosilloi 282
Helleborine flore papilionaceo, see *Anacheilium papilio*
Helleborine graminea..., see *Coilostylis ciliaris*
Holochila 149, 151
Hormidium 23, 24, 152–153, 207, 217, 251
Hormidium allemanii, see *Anacheilium allemanii*
Hormidium allemanoides, see *Anacheilium allemanoides*
Hormidium baculus, see *Anacheilium baculus*
Hormidium boothianum, see *Prosthechea boothiana*
Hormidium brassavolae, see *Panarica brassavolae*
Hormidium caetense, see *Anacheilium caetense*

Hormidium calamarium see *Anacheilium calamarium*
Hormidium chacaoense, see *Anacheilium chacaoense*
Hormidium chimborazoense, see *Anacheilium chimborazoense*
Hormidium chondylobulbon, see *Anacheilium chondylobulbon*
Hormidium coriaceum, see *Anacheilium crassilabium*
Hormidium cyanocolumna, see *Oestlundia cyanocolumna*
Hormidium cyanocolumnum, see *Oestlundia cyanocolumna*
Hormidium faresianum, see *Anacheilium faresianum*
Hormidium faustum, see *Anacheilium faustum*
Hormidium fragrans, see *Anacheilium fragrans*
Hormidium glumaceum, see *Anacheilium glumaceum*
Hormidium guttatum, see *Prosthechea guttata*
Hormidium humile, see *Hormidium pygmaeum*
Hormidium ionocentrum, see *Panarica ionocentra*
Hormidium lineatum, see *Anacheilium crassilabium*
Hormidium lividum, see *Pollardia livida*
Hormidium miserum, see *Epidendrum miserum*
Hormidium moojenii, see *Anacheilium moojenii*
Hormidium panthera, see *Prosthechea panthera*
Hormidium papilio, see *Anacheilium papilio*
Hormidium pseudopygmaeum 24, **154–155**
Hormidium pygmaeum 24, 26, 152, 153, **155–157**, 158, *185*
Hormidium racemiferum 157–158
Hormidium rhynchophorum **158–160**, *186*
Hormidium sessiliflorum, see *Anacheilium sessiliflorum*
Hormidium sima, see *Anacheilium simum*
Hormidium simum, see *Anacheilium simum*
Hormidium spondiadum, see *Anacheilium spondiadum*
Hormidium tigrinum, see *Anacheilium tigrinum*
Hormidium tripterum, see *Hormidium pygmaeum*
Hormidium uniflorum, see *Hormidium pygmaeum*
Hormidium variegatum, see *Anacheilium crassilabium*

Hormidium virens, see *Anacheilium crassilabium*
Hormidium widgrenii, see *Anacheilium widgrenii*
Hymenochila 149

Kalopternix 15, **282**
Kalopternix deltoglossus 282
Kalopternix mantinianus 282
Kalopternix sophronites 282

Laelia 15
Laelia domingensis, see *Anacheilium papilio*
Laeliacattleya Elegans 16
Laeliinae 20
Laeliopsis 150
Lanium **283**

Microstylis humilis, see *Hormidium pygmaeum*

Nidema 15, **283**
Nidema boothii 151, 283
Nidema ottonis 283

Oestlundia 23, 24, 201–202, 277
Oestlundia cyanocolumna 24, *187*, **202–203**
Oestlundia distantiflora 151, **203–204**
Oestlundia luteorosea 24, *187*, **204–205**, 206
Oestlundia tenuissima 24, *187*, **205–206**
Oncidiinae 20
Oncidium 18
Ophrys squamatum, see *Dipodium squamatum*
Osmophytum 23, 27, 28, 30, 31

Panarica 24, 207–208
Panarica brassavolae 151, *188*, 207, **208–210**, 212
Panarica ionocentra 24, *188*, **211**
Panarica mulasii 212
Panarica neglecta **212–213**
Panarica prismatocarpa 24, 26, *189, 190*, 211, **213–214**
Panarica tardiflora **214–215**
Phaedrosanthus ciliaris, see *Coilostylis ciliaris*
Phaedrosanthus cochleatus, see *Anacheilium cochleatum*
Pleuranthium 29
Pollardia 26, 151, 216–218
Pollardia campylostalix *190*, **220–221**, 263
Pollardia concolor *190*, **222–224**, 241
Pollardia ghiesbreghtiana *191*, **224–226**, 228
Pollardia greenwoodiana **226–227**
Pollardia hastata 151, *191*, 225, **228–229**, 237

Pollardia linkiana 26, 151, *191*, **229–230**, 250
Pollardia livida *192*, **231–233**
Pollardia michuacana *193*, 223, **233–235**, 241
Pollardia obpiribulbon *193*, **235–237**
Pollardia pringlei *193*, 228, **237–238**, 240
Pollardia pterocarpa 151, *194*, **238–240**, 250
Pollardia punctulata **240–242**
Pollardia semiaptera *194*, 226, **242–243**
Pollardia tripunctata 151, *194*, **244–245**
Pollardia varicosa *195*, **245–247**
Pollardia venosa *195*, *196*, **247–249**, 250
Prosthechea 23, 24, 26, 27 29, 111, 151, 207, 216, 217, 251–253
Prosthechea abbreviata, see *Anacheilium abbreviatum*
Prosthechea aemula, see *Anacheilium aemulum*
Prosthechea alagoensis, see *Anacheilium alagoense*
Prosthechea allemanii, see *Anacheilium allemanii*
Prosthechea allemanoides, see *Anacheilium allemanoides*
Prosthechea aloisii, see *Anacheilium aloisii*
Prosthechea arminii 202, 252, **254**
Prosthechea baculus, see *Anacheilium baculus*
Prosthechea bennettii, see *Anacheilium bennettii*
Prosthechea bicamerata *196*, **255**
Prosthechea bidentata, see *Prosthechea boothiana*
Prosthechea boothiana *196*, 216, 217, 252, **256–258**, 270
Prosthechea boothiana var. *erythronioides*, see *Prosthechea boothiana*
Prosthechea brachiata *197*, **259**
Prosthechea brachychila, see *Anacheilium brachychilum*
Prosthechea brassavolae, see *Panarica brassavolae*
Prosthechea bulbosa, see *Anacheilium bulbosum*
Prosthechea caetensis, see *Anacheilium caetense*
Prosthechea calamaria, see *Anacheilium calamarium*
Prosthechea campos-portoi, see *Anacheilium campos-portoi*
Prosthechea campylostalix, see *Pollardia campylostalix*
Prosthechea carrii, see *Anacheilium carrii*
Prosthechea chacaoensis, see *Anacheilium chacaoense*
Prosthechea chimborazoensis, see *Anacheilium chimborazoense*
Prosthechea chondylobulbon, see *Anacheilium chondylobulbon*
Prosthechea christii, see *Anacheilium crassilabium*
Prosthechea christyana 252, **260–261**
Prosthechea cochleata, see *Anacheilium cochleatum*
Prosthechea cochleata var. *triandra*, see *Anacheilium cochleatum*
Prosthechea concolor, see *Pollardia concolor*
Prosthechea crassilabia, see *Anacheilium crassilabium*
Prosthechea cretacea **261–262**
Prosthechea faresiana, see *Anacheilium faresianum*
Prosthechea farfanii, see *Anacheilium farfanii*
Prosthechea fausta, see *Anacheilium faustum*
Prosthechea fortunae 252, **262–263**
Prosthechea fragrans, see *Anacheilium fragrans*
Prosthechea fuscata, see *Anacheilium fuscum*
Prosthechea garciana, see *Anacheilium garcianum*
Prosthechea ghiesbreghtiana, see *Pollardia ghiesbreghtiana*
Prosthechea gilbertoi, see *Anacheilium gilbertoi*
Prosthechea glauca 24, 151, *197*, 251, 252, **263–265**
Prosthechea glumacea, see *Anacheilium glumaceum*
Prosthechea grammatoglossa *197*, 202, 252, **265–267**
Prosthechea greenwoodiana, see *Pollardia greenwoodiana*
Prosthechea guttata *198*, 252, **267–269**
Prosthechea hajekii, see *Anacheilium hajekii*
Prosthechea hartwegii, see *Anacheilium hartwegii*
Prosthechea hastata, see *Pollardia hastata*
Prosthechea inversa, see *Anacheilium bulbosum*
Prosthechea ionocentra, see *Panarica ionocentra*
Prosthechea ionophlebia, see *Anacheilium ionophlebium*
Prosthechea jauana, see *Anacheilium jauanum*
Prosthechea joaquingarciana, see *Anacheilium joaquingarcianum*
Prosthechea kautskyi, see *Anacheilium kautskyi*
Prosthechea lambda, see *Anacheilium lambda*
Prosthechea leopardina, see *Anacheilium crassilabium*

Prosthechea lindenii, see *Anacheilium lindenii*
Prosthechea linkiana, see *Pollardia linkiana*
Prosthechea livida, see *Pollardia livida*
Prosthechea maculosa, see *P. guttata*
Prosthechea magnispatha 198, **269–270**
Prosthechea megahybos, see *P. christyana*
Prosthechea michuacana, see *Pollardia michuacana*
Prosthechea moojenii, see *Anacheilium moojenii*
Prosthechea neglecta, see *Panarica neglecta*
Prosthechea neurosa, see *Anacheilium neurosum*
Prosthechea obpiribulbon, see *Pollardia obpiribulbon*
Prosthechea ochracea 24, 26, 151, 198, 252, 269, **270–272**
Prosthechea ortizii **272–273**
Prosthechea pamplonensis, see *Anacheilium pamplonense*
Prosthechea panthera 199, 252, **273–274**
Prosthechea papilio, see *Anacheilium papilio*
Prosthechea pastoris, see *Encyclia pastoris*
Prosthechea pipio see *Anacheilium calamarium*
Prosthechea pringlei, see *Pollardia pringlei*
Prosthechea prismatocarpa, see *Panarica prismatocarpa*
Prosthechea pseudopygmaea, see *Hormidium pseudopygmaeum*
Prosthechea pterocarpa, see *Pollardia pterocarpa*
Prosthechea pulcherrima, see *Epidendrum pulcherrimum*
Prosthechea pulchra, see *Anacheilium vinaceum*
Prosthechea punctifera see *Anacheilium calamarium*
Prosthechea pygmaea, see *Hormidium pygmaeum*
Prosthechea racemifera, see *Hormidium racemiferum*
Prosthechea radiata, see *Anacheilium radiatum*
Prosthechea regnelliana, see *Anacheilium regnellianum*
Prosthechea rhombilabia, see *Pollardia rhombilabia*
Prosthechea rhynchophora, see *Hormidium rhynchophorum*
Prosthechea sceptra, see *Anacheilium sceptrum*
Prosthechea semiaptera, see *Pollardia semiaptera*
Prosthechea serpentilingua 252, **274–275**
Prosthechea sessiliflora, see *Anacheilium sessiliflorum*
Prosthechea sima, see *Anacheilium simum*
Prosthechea spondiada, see *Anacheilium spondiadum*
Prosthechea squamata, see *P. christyana*
Prosthechea suzanensis, see *Anacheilium suzanense*
Prosthechea tardiflora, see *Panarica tardiflora*
Prosthechea tigrina, see *Anacheilium tigrinum*
Prosthechea tripunctata, see *Pollardia tripunctata*
Prosthechea trulla, see *Anacheilium trulla*
Prosthechea vagans, see *Anacheilium vagans*
Prosthechea varicosa, see *Pollardia varicosa*
Prosthechea vasquezii, see *Anacheilium vasquezii*
Prosthechea venezuelana, see *Anacheilium venezuelanum*
Prosthechea venosa, see *Pollardia venosa*
Prosthechea vespa, see *Anacheilium crassilabium*
Prosthechea vinacea, see *Anacheilium vinaceum*
Prosthechea vitellina 24, 26, 151, 199, 201, 202, 252, **275–277**
Prosthechea widgrenii, see *Anacheilium widgrenii*
Psilanthemum 281

Sarcochila 149, 151
Schomburgkia 150
Seraphyta 29, **283–284**
Seraphyta diffusa 283–284
Seraphyta multiflora, see *S. diffusa*
Sphaerochila 149, 151